GLOBAL WARMING AND CLIMATOLOGY

GLOBAL WARMING AND CLIMATOLOGY

Chanchal Singh

AKANSHA PUBLISHING HOUSE
NEW DELHI - 110 002 (INDIA)

AKANSHA PUBLISHING HOUSE
4649/21-B, Ansari Road
Daryaganj, New Delhi - 110 002
Phones: 23263193, 9811582579
E-mail: ektabooks@yahoo.com

Global Warming and Climatology

First Published, 2007

ISBN 978-81-8370-121-1

PRINTED IN INDIA

Published by Akansha Publishing House, New Delhi 110 002 and Printed at Capital Offset Press, Delhi

CONTENTS

PREFACE

"All progress and growth is a matter of change, but change must be growth within our social and government concepts if it should not destroy them."

—*Herbert Hoover*

"Nature has enough for man's need, but not for his greed."

—*Mahatma Gandhi*

'Global warming' and 'Climate change' are much debated terms these days which have attracted the inclination of scientists, governments, NGOs and common people world over.

The scientific community has reached a strong consensus regarding the science of global climate change. The world is undoubtedly warming. This warming is largely the result of emissions of carbon dioxide and other greenhouse gases from human activities including industrial processes, fossil fuel combustion, and changes in land use, such as deforestation. Continuation of historical trends of greenhouse gas emissions will result in additional warming over the 21st century, with current projections of a global increase of 2.5°F to 10.4°F by 2100. This warming will have real consequences for the United States (who contributes maximum greenhouse gases emission in atmosphere) and the world, for with that warming will also come additional sea-level rise that will gradually inundate coastal areas, changes in precipitation patterns, increased risk of droughts and floods, threats to biodiversity, and a number of potential challenges for public health.

It is significant to note here that addressing climate change is no simple task. To protect ourselves, our economy, and our land from the adverse effects of climate change, we must ultimately dramatically reduce emissions of carbon dioxide and other greenhouse gases. To achieve this goal we must fundamentally transform the way we power our global economy, shifting away from a century's legacy of unrestrained fossil fuel use and its associated emissions in pursuit of more efficient and renewable sources of energy.

As a natural phenomenon, global warming and consequent climate change cannot be considered as some strnge alarming happenings. But the human-induced causes have alarmingly accelerated the process. It is being feared, on melting the polar ice-sheets the sea level will rise to a dangerous heights. All low-lying coastal areas, which are currently densely populated and urbanised, will submerge. Biodiversity will come to a stake. Many species of fauna and flora found in these areas will vanish for ever. Thus in macro perspective the survival of life (floral and faunal both) on the planet earth will be at the verge of extinction.

Coming to India, due to perceptible results of climate change, we have been noticing strange things. The Thar desert of Rajasthan is getting undreamt heavy rainfalls. Barmer and Jodhpur regions are coming to the grip of extraordinary floods. Mumbai and other areas are being flooded with heavy rains. While in northern regions, scanty rainfall/snowfall is being noticed, hence visited by frequent droughts.

Though entire international community is seriously pondering over the situation, the major culprits of global warming are yet reluctant to curb/limit their hazardous activities. Here the time has come when all of us should think of the concept of `sustainability'. We have to follow the principles of sustainable development, sustainable tourism, sustainable agriculture and sustainable everything. Otherwise the posterity will be substantially in danger.

The present volume is an attempt in the direction of bringing together all necessary information on global warming and climate change, which is much debated these days. The judiciously selected readings are arranged and highlighted in coherent way. The valuable information is gathered under the following headings: Global Warming: An Overview; Greenhouse Effect: An Introduction; Climate Change: What & Why?; Climate Change in Asia; Climate Change: Awareness & International Action; Global Warming and Renewable Energy etc. The text of some statutes, rich glossary of relevant terms and bibliography are added features of this volume.

Though every care is taken to present the authentic facts and figures, any discrepancy which might have crept in inadevertently will be rectified if brought to the notice.

A work of this nature cannot be produced without extracting the findings of many experts working in the field. We have heavily drawn upon the works of various scientists, international organisations actively involved with the concerned problem.

As a famous adage of E.W. Wilcox goes on—"he who compiles a book of helpful philosophy out of the material provided by other minds does the world a greater service than he who creates an epic of despair." Here we owe a deep sense of gratitude to the original contributors on whose works, this book is devised and developed. The entire credit ought go to them all.

We acknowledge our gratefulness to one and all who have guided, inspired, encouraged and assisted us while working on this project. Finally, we express our thankfulness to our publisher for evincing keen interest in this and executing the entire work superbly.

—CS

INTRODUCTION

Since the impact of 'global warming' and 'climate change' on traditional life style of people and other living beings in all regions has become perceptible more or less. Alarmed by this almost all governments have started seriously considering over the problem and steps are being taken to avert the causes of global warming as much as possible.

In case of India, as disseminated through media, the Prime Minister, Dr Manmohan Singh, has set up a high level advisory group on climate change issues to coordinate national action plans for assessment, adaptation and mitigation of climate change. It will advise the Government on pro-active measures that can be taken by India to deal with the challenge of climate change. It will also facilitate inter-ministerial coordination and guide policy in relevant areas.

Stressing on the need to protect the environment while ensuring higher growth rates, Dr. Manmohan Singh stated: "The threat of climate change is real and unless we alter our lifestyles and pursue a sustainable model of development, our future will be at peril."

Comong to the composition of Prime Minister's Council on Climate Change, include the Minister of External Affairs; Finance Minister; Minister of Environment & Forests; Minister of Agriculture; Minister of Water Resources; Minister of Science & Technology; Deputy Chairman,

Planning Commission; Dr. R. Chidambaram, Principal Scientific Advisor to PM; Mr V. Krishnamurthy, Chairman, NMCC; Dr. C. Rangarajan, Chairman, EAC; Mr Ajay Mathur, Chairman, Bureau of Energy Efficiency; Foreign Secretary; Secretary, Environment & Forests and Principal Secretary to PM (Convenor).

Further, it would also have non-Government Members like Dr R.K. Pachauri, Chairperson, TERI; Dr Prodipto Ghosh; Dr Nitin Desai; Dr Sunita Narain; Mr Chandrashekhar Dasgupta (Retd. Diplomat); Mr Ratan Tata, Chairman, Investment Commission; Mr Raj Chengappa, Executive Editor, *India Today*; Dr R. Ramachandran, Science Editor, *The Hindu*.

On July 13, 2007 Prime Minister of India chaired the first meeting of National Council on Climate Change. Taking the first step towards developing a national plan to tackle the effects of global warming and cut back greenhouse gas emissions, Prime Minister Manmohan Singh said the government would convert six million hectares of degraded forest land into green areas.

A major afforestation programme called 'Green India', which will kick off on Independence Day, was announced by Manmohan Singh at the first meeting of the National Council on Climate Change set up in June 2007.

With reports that receding Himalayan glaciers could jeopardise water supplies for millions of people in the country, he also called for a long-term strategy to deal with glacial melting. "Our food security comes largely from irrigated areas of Punjab, Haryana and Uttar Pradesh whose rivers are fed by glacier melting in the Himalayas. There is a gap in our understanding of the Himalayas and we need to build a knowledge-based partnership of affected countries to manage and develop the region," said Manmohan Singh.

"As a concrete first step, we must identify knowledge institutions within the country that we can mandate for an agenda related to the Himalayas as well as identify other knowledge institutions with whom we could collaborate in knowledge-sharing."

The meeting decided that a national strategy paper on climate change would be prepared before the end of this year. R.K. Pachauri, chairperson of TERI and a member of the council, has said that India was completely unprepared for the effects of climate change that could lead to widespread flooding and drought.

"On a scale of 0 to 10, India's preparedness is 0.5," Pachauri had warned at the United Nations Intergovernmental Panel on Climate Change (IPCC) in April.

In his opening remarks, Manmohan Singh said India was in the forefront of development of the Clean Development Mechanism (CDM) and the largest global player in terms of the number of projects. "The Indian National CDM Authority has accorded host country approval to over 667 projects facilitating an investment of nearly Rs.600 billion in sectors such as energy efficiency, fuel, industrial processes, municipal solid waste and renewable energy," the prime minister said.

Participants at the meeting emphasized the need for funding research on impact of climate change including research on management of impact of droughts and floods on crop production and urban planning.

Some of the other issues discussed were encouraging public transportation in urban areas and reducing dependence on fuel inefficient technologies.

The setting up of the panel is thought to be a response to consistent demands that both India and China — likely to be the world's largest polluters in the future — commit to major cuts in carbon emissions.

Global warming was on top of the agenda at the G-8 meeting at the German spa town of Heiliegendamm

However, India and other developing nations contend that the developed world must subsidise clean technologies in order to allow them to make the switch to less-polluting energy sources without compromising on economic growth.

The Prime Minister directed the Planning Commission to incorporate clean development strategies into the sectoral plans and proposals for the Eleventh Plan and to make a strategy to deal with climate change an intrinsic part of the Eleventh Plan strategy.

1

GLOBAL WARMING: AN OVERVIEW

Do not save your loving speeches
For your friends till they are dead;
Do not write them on their tombstone.
Speak them rather now instead.

—Anna Cummins

By 'Global Warming' we mean an increase in the Earth's temperature. The earth is gradually changing from cold to hot. Since the Industrial Revolution the Earth has been getting warmer. The average temperature has gone up 1°F for the past 100 years. This may seem like a little bit but, since it's global, it's huge. If the Earth's temperature rises more than 1°F over the next few years, the effects will be catastrophic. The Earth will flood and the wildlife, plants, and humans will be gone. The Earth will have no life in it. Some climatologists are trying to find a way to reduce Global Warming without having to ban factories because they are necessities. Factories produce our clothes and many other products that we need. Banning them would decrease Global Warming, but people would have to make their own necessities. Global warming poses an extraordinary challenge. The world's leading atmospheric scientists tell us that a gradual warming of our climate is underway and will continue. This long-term warming trend poses serious risks to our economy and our environment. It poses even greater risks to many other nations, particularly poorer countries that will be far less able to cope with a changing climate and low-lying countries where sea level rise will cause significant damage.

Meeting the challenge of global warming will require sustained effort over decades—on the part of governments, who must establish the rules and modify them as we learn more of the science, and as technological solutions begin to manifest themselves; on the part of industry, who must innovate, manufacture, and operate under a new paradigm where climate change will drive many decisions; and on the part of the public, who must also switch to a more climate-friendly path in their purchases and life-styles.

Basic Facts

The scientific community has reached a strong consensus regarding the science of global climate change. The world is undoubtedly warming. This warming is largely the result of emissions of carbon dioxide and other greenhouse gases from human activities including industrial processes, fossil fuel combustion, and changes in land use, such as deforestation. Continuation of historical trends of greenhouse gas emissions will result in additional warming over the 21st century, with current projections of a global increase of 2.5°F to 10.4°F by 2100, with warming in the U.S. expected to be even higher. This warming will have real consequences for the United States and the world, for with that warming will also come additional sea-level rise that will gradually inundate coastal areas, changes in precipitation patterns, increased risk of droughts and floods, threats to biodiversity, and a number of potential challenges for public health.

Addressing climate change is no simple task. To protect ourselves, our economy, and our land from the adverse effects of climate change, we must ultimately dramatically reduce emissions of carbon dioxide and other greenhouse gases. To achieve this goal we must fundamentally transform the way we power our global economy, shifting away from a century's legacy of unrestrained fossil fuel use and its associated emissions in pursuit of more efficient and renewable sources of energy.

Such a transformation will require society to engage in a concerted effort, over the near and long-term, to seek out opportunities and design actions to reduce greenhouse gas emissions. Global warming is caused by a blanket of pollution that traps heat around the earth. This pollution comes from cars, factories, homes, and power plants that burn fossil fuels such as oil, coal, and gasoline.

Air is all around us. Air supplies us with oxygen to breath. Air pollution includes smog, carbon dioxide, methane, as well as other polluted gases. For example diesel smoke is the black gas that comes from burning fuels in cars, homes, and industries. Another type of pollution is noxious gases which are sulfur dioxide, carbon monoxide and nitrogen oxides. Pollution inside our homes includes tobacco smoke, cooking, and heating appliances. In order to control air pollution in our home, school, and work place, we need to have adequate ventilation.

Causes of Global Warming

Global warming is the direct result of climate temperatures rising, pollution and deforestation. During the past decades scientists have noticed that the temperatures in the sea level are rising. The hot summers are getting longer and our cold winters are getting shorter duration. Although we are not to blame and have no control over the climate, we can control the amount of pollution that we make and stop deforestation.

Animals and Global Warming

Global warming has affected wildlife all around the world in many ways. Some animal species are predicted to die out or become endangered. The Earth's temperature is changing very fast and wildlife will not be able to adapt to the changes this fast. Adapting to these changing temperatures can take wildlife millions of years. Frogs, Polar Bears, Penguins, Monarch Butterflies, Cold Water Fish, Coral Reefs and Arctic Foxes many Himalayan species of are some of the many animals affected by global warming.

Harlequin Frogs are dying because of global warming. A very deadly fungus is growing on them. The atmosphere is heating up and that is making the fungus grow. This fungus makes some frogs lose their legs or grow more legs.

Polar Bears depend on sea ice. But the ice is melting. Scientists blame global warming for this. If the sea ice melts, then the Polar Bears will not have a habitat to live in.

Similarly, the Adeline penguins better known as tuxedoed penguins are in danger because their habitat is getting warmer. Warmer weather causes the amount of algae in the water to decrease the krill shrimp. Krill shrimp are the main food source of the Adeline penguins. The change in climate has also effected the monarch butterflies. Their migration patterns of Canada and North America have changed lately. Now they have been migrating to Mexico in order to protect themselves from rain and freezing temperatures. Cold water fish are loosing their habitat as the temperatures rise. Cold water fish need their water to be a frigid mixture of spring and glacier water.

Coral reefs are decreasing in the ocean as the levels of carbon dioxide increases in the water. Coral reefs play an important role in ocean as they act as a natural protective barrier for the coastal regions. Arctic Foxes are also looking for a new colder home. The increase of temperatures are pushing this thick furry animals to look for cooler places or otherwise extincts.

Supported by scientific evidence, it is now obviously accepted that the earth's climate is changing. The atmosphere is warming and this trend will continue. Because of the complexity of atmospheric and oceanic currents this warming will produce violent storms, drought and floods and other not-yet-predictable weather events. Already several examples of atmospheric warming are available from around the globe. For example, nine of the hottest years recorded in more than a century have occurred since 1988. First, July 1998 was the hottest month ever; second, in 1998 India experienced its worst hot spell in 50 years, which took a toll of over 3,000 lives and thirdly, another startling phenomenon is the retreat of the Himalayan Glaciers—18 m per year in the case of Gangotri. These changes are happening because humans have released various heat-absorbing gases into the atmosphere—mainly through burning fossil fuels e.g. coal and oil.

For the past two decades scientists have been collecting and debating evidence of long-term climate change. Are the observed warming trends simply natural variations in climate or are they a long- term trend? And if there is a trend what is causing it—human activity or natural fluctuations? In 1988 the United Nations set up the

Intergovernmental Panel on Climate Change (IPCC)—an official scientific body comprised of leading atmospheric scientists to investigate climate change. IPCC's Second Assessment Report published in 1995 states that climate change is a long-term trend, and human activities are its major cause. At the root of this is the human use of fossil fuels. When burnt they release what are called greenhouse gases (GHGs).

But, the release of main GHGs—carbon dioxide, methane, and nitrous oxide—is not only due to the burning of fossil fuels. It is also a part of nature's normal processes. Like the panes of a greenhouse, GHGs allow sunlight to pass through the troposphere (lower atmosphere), but trap the heat. As the heat rises from the earth's surface, into the troposphere, some of this heat escapes into space, some is reflected back to the surface by the molecules of GHGs, warming the air.

This natural trapping of heat, or the greenhouse effect, has made earth habitable. Without it the earth would have been a cold, lifeless planet. As such, in the normal scheme of things GHGs, which make up less than 1 per cent of the atmosphere, are benign. Their levels in the atmosphere are determined by a balance between "*sources*" (processes which release these gases) and "*sinks*" (processes which absorb or remove them). But a lot of modern human activity tends to disrupt this optimal balance.

Such disruption may happen by way of introducing new or additional sources of natural GHGs, or manmade GHGs such as CFCs and their substitutes, or of interference with natural sinks. The enhanced levels of GHG accumulation in the atmosphere resulting from this disruption are the cause of Global Warming and Climate Change.

Remarkably, Greenhouse gases levels are rising as a direct result of human activity. Additions by human activity can significantly affect the amount of heat trapped in the atmosphere over time, and most of these gases have fairly long life spans ranging from ten to thousands of years. Notably, what we put into the atmosphere today will therefore continue to warm the planet for a long duration in future.

Potent Warmers of Globe

First and foremost, **Carbon dioxide** (CO_2) is responsible for over 60 per cent of the current global warming from GHGs produced by human activities since preindustrial times (around 1750). Its concentration since then has increased by more than 30 per cent and currently increases by 1 per cent every year. The main sources (75 per cent) are the burning of fossil fuels, particularly coal, and increasingly, motor vehicle exhaust. Deforestation and biomass burning contribute 25 per cent. CO_2 remains in the atmosphere for around 200 years.

Secondly, **Methane** (CH_4) can trap 20 to 25 times more heat than CO_2 on a molecule for molecule basis. It is produced by decomposition of organic matter in rice fields, natural wetlands, landfills, intestines of cattle, sheep and termites, and in natural gas leaks. Its concentration has doubled since preindustrial times. It stays in the atmosphere for only 10-12 years, but it is removed when it reacts with the hydroxyl (OH) radical to form CO_2.

Further **Nitrous oxide** (N_2O) can trap 200 times more heat than CO_2 on a molecule for molecule basis and has a life span of 120 to 190 years. It is released from a variety of sources including the burning of biomass and coal, the application of nitrogen fertilizers, and from nylon production.

Its concentration is growing at the rate of 0.25 per cent per year.

Again, **Chlorofluorocarbons** (CFCs) can trap 1,500 to 1,700 times more heat than CO_2 on a molecule for molecule basis and remain in the atmosphere for several thousand years. The main sources of CFCs are leaking refrigerants, industrial solvents, aerosol propellants and production of plastic foams. Their concentration had been growing by 4 per cent per year in the 1990s, but their use is now being phased out under the Montreal Protocol because of their ozone-depleting properties. As a result, CFCs are not included within the scope of the Framework Convention or the Kyoto Protocol. The substitutes developed for CFCs do not directly destroy ozone in the earth's atmosphere but do contribute to global warming. Hydroflurocarbon gases (HFCs) are a manmade alternative for CFCs for use in refrigeration, as agents used to blow foams or insulation, and as solvents or cleaning agents specially in semiconductor manufacturing. Their global warming potential is, however, 4,000 to 10,000 times that of carbondioxide.

While **Perfluorocarbons** (PFCs) are replacement gases for CFCs, but are also a by-product of aluminium smelting. Small amounts are also produced during the uranium enrichment process. They can trap 6,000 to 10,000 times more heat than CO_2 as GHGs. Sulphur hexafluoride (SF_6) is a manmade gas used as insulating material for high-voltage equipment such as circuit breakers. It is also used for detecting water leaks in cable cooling systems. It can trap 25,000 times more heat than CO_2.

Although currently, the CFC substitutes listed above contribute little to global climate change, the projected growth in their use could contribute to it significantly in the 21st century.

Concentration of greenhouse gases in the atmosphere has continued to increase. Atmospheric concentrations of CO_2, CH_4 and N_2O have increased by 30 per cent, 145 per cent and 15 per cent respectively since preindustrial times. CO_2 concentration, for example, has increased from 280 ppmv (parts per million by volume) in the 1750s to almost 360 ppmv in 2000. The IPCC has set 450 ppmv as the top concentration at which greenhouse gases should be stabilized by the end of the 21st century.

Effects

Most scientists agree that the earth's mean temperature has risen by at least 0.6ºC over the last 120 years. Global warming will lead to a rise in mean sea levels as water expands when heated. The earth's average sea level is expected to rise by about 50 cm by 2100, flooding many low-lying islands and coastal areas. Some of them, such as many islands of the Maldives and large parts of southern Bangladesh, may even become permanently submerged. Sunderban delta (India) is also at the brink of extinction. The energy imbalance in the climate system caused by global warming will result in more violent weather events, increasing the threat of drought and floods (already a major factor due to other kinds of environmental degradation) and intense storms. Global warming also poses serious threats to food production, fresh water sources and human health.

Tropical diseases such as malaria could spread to formerly temperate zones, affecting 60 per cent of humanity. Sea level rise (SLR) and changing weather patterns could trigger large-scale migration from more seriously affected areas. By 2050 global warming could produce as many as 150 million environmental refugees, most of whom would migrate to other countries, causing social tensions and political instability. All these consequences of global warming will translate into huge financial costs.

Scientist estimate that just to stabilize the levels of CO_2 in the atmosphere at their current levels, current global emissions will have to be cut by 60 to 80 percent. But projections suggest that between 1990 and 2010 CO_2 emissions will increase by nearly 50 per cent. Recognizing climate change as a serious global problem, countries of the world got together to discuss what could be done to check the threat. The First World Climate Conference was held in 1979. A formal system of governmental negotiations to tackle the problem was subsequently initiated which resulted in the UN Framework Convention on Climate Change. The legally-binding Kyoto Protocol requiring industrialized countries to make GHG emission reductions is currently open for ratification.

Global Warming and the Weather

Admittedly, during the past years we have experienced heat waves. Warmer weather is causing the glaciers to melt around the world. Warmer temperatures also speeds up the process of evaporation which leads to droughts. Recently global warming was felt in Louisiana by the Hurricane Katrina. This happen because warmer weather is melting away the ice glaciers making more water. More water plus hurricanes equal risen sea levels which means problems for those who live close to the coast. The imbalance in weather will have a negative impact in humans, animals and plants.

Projected Effects

- The earths temperture will rise several degrees in the next century.
- An increase in precipitation.

- More intense rainstorms.
- Drier soils in specific areas.
- Warming will occour faster over land than over water.

Global Warming, and Sea Level Rise (SLR)

The warming temperature will raise sea level because the melting glaciers and parts of Greenland Ice Sheet.

Two primary processes contribute to sea level rise (SLR):

1. Thermal expansion—the increase in water volume resulting from heat uptake;
2. Mass inputs—the transfer of freshwater from land to the ocean. Large ice sheets on Greenland and Antarctica represent the largest potential source of additional water mass.

There are two general mechanisms by which such ice sheets transfer water to the ocean:

1. Surface melting of ice on land with subsequent runoff into the ocean;
2. Ice dynamics, wherein ice actually flows from land into the ocean

New evidence indicates that ice flow is accelerating around Greenland and Antarctica, making this process the greatest source of uncertainty for predicting future SLR as warming proceeds.

During Ancient Period

Over the past four cycles, each lasting about 100,000 years, sea level rose and fell by about 400 feet, with ice ages having lower sea level and warm interglacial periods having higher sea level. Contributions from land-based ice sheets dominated SLR during interglacial warming.

During the warmest part of the last interglacial period (about 130,000 years ago), global average temperature was 2-3 °F warmer than today and global sea level was 13-20 feet higher. During the Middle Pliocene (3 million years ago), global temperature was 3.5-5.5 °F warmer than today and sea level was 80-115 feet higher. If emissions of man-made greenhouse gases continue without abatement, the earth could warm by 5.5 °F (3 °C) within the next century.

The complete loss of large ice sheets would take from several centuries to millennia, but warming could cross a threshold within decades that could permanently destabilize large ice sheets.

Current Status

Based on tide gauges, sea level rose by an average of 0.7 inches per decade and accelerated by 0.05 inches per decade over the 20th century.

More accurate satellite measurements indicate that global sea level has risen by 1.2 inches over the past decade, about 70% faster than the 20th century average.

Most of the scientists assumed that thermal expansion dominated contemporary SLR, but recent progress reveals that freshwater contributions from land dominate, consistent with recent acceleration of ice loss from glaciers.

Further Projection

Because the ocean has an enormous thermal inertia, it takes many decades for sea level to adjust to a quantity of heat that it absorbs. This delay means that even if man-made greenhouse gas emissions were halted today, sea level rise would be committed to an additional 0.4-1.0 foot by the end of this century. In 2001 the Third Assessment Report of the Intergovernmental Panel on Climate Change (IPCC) projected that sea level would rise by 0.3-2.9 feet by the end of the 21st century.

In 2007 the Fourth Assessment Report of the IPCC projected that sea level would rise by 0.6-1.9 feet by the end of the 21st century.

The lower ends of the projected SLR ranges from the Third and Fourth IPCC reports are similar to the sea level rise commitment described above, leaving little or no room for additional SLR generated by continued greenhouse gas emissions. The upper end of both omits the uncertainty associated with the future ice dynamical changes in Greenland and West Antarctica.

Current SLR models undershoot the sea level change observed during the 20th century, possibly because they do not adequately simulate freshwater mass contributions from land. As an alternative, a recent study extended the statistical relationship between temperature rise and SLR observed during the 20th century, using models of future temperature change to drive SLR. This empirical approach estimated 21st century SLR to be in the range of 1.6-4.5 feet, if manmade greenhouse gas emissions continue to grow.

Both long-term tide gauge measurements and recent satellite measurements suggest that SLR accelerated during the 20th century, concomitant with increasing global temperature.

One of the most significant developments of recent years is the realization that mass contributions from land ice have dominated contemporary SLR.

Large polar ice sheets are more sensitive to surface warming than previously realized, and large dynamical changes are now being observed on the Greenland and West Antarctic ice sheets.

New understanding of how these ice sheets behaved in the past suggests that they could add water mass to the oceans much more quickly than previously assumed.

The Fourth Assessment Report of the IPCC projects a 21st century sea level rise of 0.6-1.9 feet by 2100, but notes that accelerated glacial melt could contribute another 0.7 feet, raising the upper end to 2.6 feet.

A new empirical projection of SLR for the 21st century indicates that sea level could rise by 1.6-4.5 feet this century.

The potential for the rapid collapse of large ice sheets seems more plausible than in the recent past, and some scientists therefore warn that continued anthropogenic warming could result in the triggering of abrupt sea level rise within the current century.

Global Warming and Agriculture

The change of global climate effects crop growth. Lately, there have been a lot of droughts, floods, and storms causing many fruits and vegetables to be uneatable. As the weather changes, there will be periods of drought followed by periods of storms and floods. If a drought happens in the early stages of the crops, it will ruin the fruits and the vegetables as well as other vegetation. Floods and storms will wash the crops away and farmers will have a hard time to grow their crops due to the poor conditions of the soil. However, some scientist believe that Global Warming will have a positive impact in agriculture because plants need carbon dioxide to grow. Other scientist believe that although carbon dioxide will help the crops to grow, we may lower the nutritional value of the crops. As of now no one is sure of the effects of Global Warming in agriculture. They do know that it will effect the growth of crops. What do you think? Take a poll.

Agriculture

Flooded Crops

Floods affect agriculture by destroying the crops. Floods also destroys wildlife habitats and increases diseases in wild animals and humans who live close to the floods.

Dry Crops

Warmer weather affects agriculture too. The plantations are much too dry, then the plants don't get enough water. Plants need water to grow. If the crops can't grow, then less and less people will be fed. The food will be more expensive and people will not be able to buy them. As a result, some people will die from starvation because of no grains, vegetables and fruits.

By the year 2050, the predicted population in the world will be 9 billion people. With the rise is temperature, the effect on growing vegetables and needed medical plants will be hindered. With every 1 degree rise in global temperature, the grains, rice and wheat growth will be nearly 10% less. In summary, the hotter our planet gets, the less the food it can produce.

The Arctic Climate Impact Assessment (ACIA) is an international project to collect and evaluate knowledge on climate variability, global warming, and increased UV radiation in the Arctic and their effects on the Arctic region, its ecosystems and communities. The final ACIA will consist of three reports: a scientific report, a synthesis report, and a policy report. The science report will detail the current state of scientific knowledge concerning climate change and its effects in the arctic. The synthesis report will be a comprehensive, but less detailed, review of the science report. The policy report will provide policy recommendations for governments working together to reduce the adverse effects of global warming.

Global Warming and Arctic Region: Main Issues

The ACIA (Arctic Climate Impact Assessment) was launched in 2000 and has been conducted as a joint initiative between the Arctic Council and the International Arctic Science Committee (IASC). The ACIA is managed through its Secretariat office, located at the International Arctic Research Center at the University of Alaska Fairbanks.

Funding has been provided by the U.S. National Science Foundation and the U.S. National Oceanic and Atmospheric Administration. More than 300 scientists and six arctic indigenous peoples' organizations have participated in preparing the reports. Another 225 additional researchers not directly affiliated with the ACIA served as peer-reviewers for the scientific and synthesis reports to ensure scientific quality and accuracy.

Arctic Council and the International Arctic Science Committee

The Arctic Council is an intergovernmental organization (members of which include Canada, Denmark, Finland, Iceland, Norway, the Russian Federation, Sweden and the United States) that collaborates to address the common concerns and challenges faced by the Arctic governments and the people of the Arctic.

Whereas the International Arctic Science Committee is a non-governmental organization that facilitates cooperation in all aspects of arctic research.

The ACIA synthesis and scientific reports was released during the ACIA International Scientific Symposium in Reykjavik, Iceland, November 9th-12th 2004. The release of the policy report has been scheduled for November 24th.

Effects of Global Warming on the Arctic

For years, climate scientists have believed that the Artic would likely be one of the first regions to be affected by global warming and would likely experience greater warming than the rest of the world. Recent evidence has validated these concerns. While the world as a whole warmed about 1°F over the entire 20th century, parts of the Arctic have warmed by 4-5°F just since the 1950s. The Arctic continues to warm at a rate about twice as fast as rest of the world.

Scientists, as well as the indigenous people of the Arctic, have noticed dramatic changes in the Arctic environment that has affected ecosystems and wildlife, human settlements and infrastructure, and the way of life of indigenous peoples. For these reasons, the ACIA was undertaken to evaluate whether these changes are caused by human activities; how Arctic climate change may affect climate change in the rest of the world; and the risks of continued global warming for the Arctic, its people, and its ecosystems and wildlife.

It is significant to note that climate change in the Arctic is expected to affect other parts of the world. The melting of ice masses in the Arctic could contribute significantly to global sea-level rise, and the addition of that fresh water to the salty oceans could change global ocean circulation patterns. Arctic tundra also stores huge amounts of carbon, which could be released to the atmosphere during a thaw, further enhancing the greenhouse effect and global warming.

Some of these impacts are detailed in a new Pew Center report, "Observed Impacts of Climate Change on Natural Systems in the United States" by Camille Parmesan and Hector Galbraith, released on November 10, 2004. Observed impacts in Alaska include the northward migration of treelines, increased melting of permafrost and the release of carbon dioxide from the thawing Tundra, and changes in competition between species such as the arctic and red fox.

Main Research Areas

- A Synthesis of Potential Climate Change Impacts on the U.S.
- Agriculture & Global Climate Change
- Aquatic Ecosystems and Global Climate Change
- Coping with Climate Change: The Role of Adaptation in the U.S.
- Coral Reefs & Global Climate Change: Potential Contributions of Climate Change to Stresses on Coral Reef Ecosystems
- Ecosystems & Global Climate Change Sea-Level Rise & Global Climate Change
- Forests & Global Climate Change: Potential Impacts on U.S. Forest Resources
- Human Health & Global Climate Change
- The Science of Climate Change
- Water and Global Climate Change

Remarkably, the rapid warming that has been observed in the Arctic region and is projected to continue well into the future is caused by a number of factors, but one of the most important is the so-called "ice-albedo feedback." It is uncertain what percentage of the observed warming is explained by the ice-albedo feedback, because there are other possible influences, such as natural variability in Arctic temperatures and the transport of heat to the Arctic by the oceans.

Besides increasing global temperatures, global warming contributes to the melting of polar ice over the Arctic Ocean and land surfaces. Such ice masses have a high albedo, meaning they reflect more incoming solar radiation than many other surfaces. As the Arctic warms and the ice melts, the uncovered land or water has a lower albedo, or a lower reflectivity. As a result, more solar radiation is absorbed at the surface, which amplifies the warming effect. The ACIA will provide a valuable scientific analysis of climate change in the Arctic region and how the Arctic interacts with the global climate.

Undoubtedly, the ACIA will call attention to the physical changes that have and are occurring in the Arctic and their effects on the Arctic's indigenous peoples as well as its natural resources. The ACIA will also be the primary assessment of possible future consequences of climate change. It explores adaptation, mitigation, and research options for limiting the adverse consequences of climate change and will help improve scientific understanding of Arctic and global climate change.

Lastly, the explicit incorporation of the knowledge and experience of indigenous peoples into the assessment process will also serve as a model for integrating stakeholder and scientific expertise in future scientific assessments across multiple disciplines.

Tropical storms form over warm tropical oceans during periods when local sea surface temperatures are above 26.5°C. Under these conditions, evaporation from the ocean surface generates very high humidity in the atmosphere, which in turn generates thunderstorms. A tropical storm forms when a system of powerful thunderstorms converges and begins to rotate in the atmosphere, forming a vortex known as a *tropical depression*.

Heat from the ocean surface is drawn up through the center of the vortex and released to the atmosphere as water vapor condenses to form rain around the perimeter of the vortex. Energy from ocean heat also generates high winds. The more heat available in the surface water, the more potential there is to generate heavy rain and high wind. If wind speeds exceed 35 mph, the National Hurricane Center deems the system a *tropical storm* and assigns a name.

Notably, a hurricane is simply an intense tropical storm with maximum sustained wind speeds of 74 mph or greater. The National Hurricane Center categorizes hurricanes according to their maximum wind speed based on the Saffir-Simpson scale.

High winds associated with hurricanes push ocean water onto the shore during land fall, causing a storm surge. The potential for damage from storm surge, heavy rain, and high wind increases dramatically as wind speeds increase, and hurricane categories indicates the potential damage from a given storm. The National Hurricane Center defines major hurricanes, those with the most destructive potential, as category 3 or higher. Note, however, that a tropical storm does not have to become a hurricane to be destructive.

For example, tropical storm Arlene made Landfall in Florida with winds of 60 mph, dropped 6 inches or more of rain in locations from Florida to New York, spawned tornadoes in Indiana, caused power outages in New York, and caused monetary damages of more than $10 million in June 8, 2005.

Remarkably the terms "tropical storm" and "hurricane" are standard in North America, but not in other parts of the world. The more general term for "tropical storm" is "tropical cyclone." In some regions of the world, hurricanes are called *typhoons* or simply tropical cyclones.

The North Atlantic and Eastern Pacific hurricane seasons, which directly affect the U.S., run from June 1-November 30, and the peak season is from mid-August through October. However, it has become more common in the past decade for hurricanes to occur outside of this window. During 2003-2005, for instance, the last named tropical storm occurred in December.

Further, here are three terms generally used to describe a hurricane season: *frequency*, *intensity* and *activity*. Hurricane frequency refers to the number of hurricanes that occur. Hurricane intensity is a measure of the strength or maximum wind speed of a hurricane (the category). Hurricane activity is the term used by the National Hurricane Center that encompasses both the frequency and intensity of hurricanes in a season. The National Hurricane Center predicted that the 2004 Atlantic hurricane season would see above normal tropical storm activity, with 12-15 named storms, 7 reaching hurricane strength. This prediction was close but conservative, as there were 15 named storms and 9 hurricanes.

The number of hurricanes making landfall in the U.S. was also above normal. Monetary damages from these events exceeded $40 billion and were concentrated in Florida, which had never before experienced more than three land-falling hurricanes in a single season.

It is remarkable to opte here that the following activity occurred during the 2005 season:

- 28 tropical storms for the first time since systematic record keeping began about 150 years ago. The previous record was 21 storms, set in 1933.
- 15 hurricanes
- 7 major hurricanes
- The earliest date on record by which four named tropical storms formed (Arlene, Bret, Cindy, Dennis formed before July 5)
- The earliest date on record by which two category 4 hurricanes occurred (Dennis formed July 4-7; Emily formed July 10-16)
- The most powerful hurricane ever recorded in the Atlantic basin (Wilma, central barometric pressure of 882 mBar)
- Three of the six most powerful hurricanes ever recorded in the Atlantic basin (Katrina, Rita, Wilma)
- The first time three category 5 hurricanes have ever been recorded in the same year in the Atlantic basin

- The most destructive hurricane in US history (Katrina)
- Tied the record for latest date of storm formation set in 1954 (tropical storm Zeta formed on Dec. 30, 2005).

In terms of overall hurricane activity, the 2005 Atlantic hurricane season was the most active season ever recorded.

Globally (not just in the North Atlantic), there is an average of about 90 tropical storms every year. This average changes very little over time and there has been no detectable change over the 20th century overall, although we do not have good historical data for all regions where hurricanes occur.

Recently on 6th June 2007, a named storm in Oman devastated the people and property in the large scale in the North Atlantic, for which we have the best records, there has been a clear increase in the number and intensity of tropical storms and major hurricanes. From 1850-1990, the overall average number of tropical storms was about 10, including about 5 hurricanes. Since 1995, the 10-year average has risen dramatically, with the 1997-2006 average at about 14 tropical storms, including about 8 hurricanes. This increase in frequency correlates strongly with the rise in North Atlantic sea surface temperature.

Of late, here is an ongoing scientific debate about the link between increased North Atlantic hurricane activity and global warming. The 2007 report of the Intergovernmental Panel on Climate Change (IPCC) rates the probability of such a link as "more likely than not".

The National Hurricane Center predicted a very active 2006 season, but the number of storms fell far short of initial predictions. With 9 tropical storms, 5 hurricanes, and 2 major hurricanes, the 2006 season looked more like a normal year based on the long term average than like its extremely active predecessors. Note that despite this low activity in the Atlantic, the 2006 hurricane season in the Pacific was very active. One likely reason for the lower Atlantic hurricane activity in 2006 is that an unanticipated El Niño event suddenly began forming during 2006. Another factor may have been large dust storms from the African continent that shaded the sea surface, lowering sea surface temperatures in the tropical North Atlantic.

The El Niño Southern Oscillation is a pattern of short-term climate variability in the tropical Pacific. Warm phases are known as El Niño events and cold phases are known as La Niña events. El Niño events in the Pacific, which occur every 4-7 years, tend to suppress hurricane activity in the Atlantic, particularly inhibiting the formation of major hurricanes. For example, every year since 1995 has seen above-average hurricane activity, with two exceptions: 2006 and 1997—both El Niño years. La Niña events create conditions more favorable for Atlantic hurricanes as a natural phenomenon.

According to the 2007 Fourth Assessment Report of the IPCC-AR4, it is "more likely than not" that there is a human contribution to the observed trend of hurricane intensification since the 1970s. In the future, "it is likely that future tropical cyclones (typhoons and hurricanes) will become more intense, with larger peak wind speeds and more heavy precipitation associated with ongoing increases of tropical.

According to the IPCC-AR4, on a global scale, "[t]here is no clear trend in the annual numbers [i.e. frequency] of tropical cyclones." However, the frequency of tropical storms has increased dramatically in the North Atlantic. Reasons for this increase are currently subject to intense debate among climate scientists.

No exaggeration, Global temperature has increased and precipitation patterns have changed over the 20th century as a result of human-induced global warming, resulting in some increases in extremes of temperature and precipitation. According to the IPCC-AR4, "[i]ncreases in the amount of precipitation are very likely in high-latitudes, while decreases are likely in most subtropical land regions," and "[i]t is very likely that hot extremes, heat waves, and heavy precipitation events will continue to become more frequent." The biggest reason for increased loss of life and property in coastal regions is population growth and increasing development in coastal areas. As growth and development continue, the damages caused by severe weather will increase regardless of global warming. It stands to reason that climate change, namely sea level rise (SLR) and increases in tropical storm activity, would exacerbate the damage as global warming continues.

II

TEMPERATURE RECORD

The temperature record shows the fluctuations of the temperature of the atmosphere and the oceans through various spans of time. The most detailed information exists since 1850, when methodical thermometer-based records began. There are numerous estimates of temperatures since the end of the Pleistocene glaciation, particularly during the current Holocene epoch. Older time periods are studied by paleoclimatology.

Instrumental Temperature Record

The Instrumental Period: From 1850

The instrumental temperature record shows the fluctuations of the temperature of the atmosphere and the oceans as measured by thermometers. A quasi-global record exists since about 1850.

The time period for which reasonably reliable near-surface temperature records exist from actual observations from thermometers with quasi-global coverage is generally considered to start in about 1850—earlier records exist, but coverage and instrument standardisation are less.

In the present day most meteorological observations are taken for use in weather forecasts. Centres such as ECMWF show instantaneous map of their coverage; or the Hadley Centre show the coverage for the average of the year 2000. Coverage for earlier in the 20th and 19th centuries would be significantly less. While temperature changes vary both in size and direction from one location to another, the numbers from different locations are combined to produce an estimate of a global average change. Most of the warming occurred during two periods: 1910 to 1945 and 1976 to 2000; the cooling/plateau from 1945 to 1976 is attributed to sulphate aerosol. Attribution of the temperature change to natural or anthropogenic factors is an important question: see global warming and attribution of recent climate change.

The data for the record come from thermometer measurements from land stations and ships, which independently show much the same warming since 1860. The data from these stations show an average surface temperature increase of about 0.74 °C during the last 100 years. The Intergovernmental Panel on Climate Change

(IPCC) stated in its Fourth Assessment Report (AR4) that the temperature rise over the 100 year period from 1906-2005 was 0.74 °C [0.56 to 0.92 °C] with a confidence interval of 90%.

The US National Academy of Science, both in its 2002 report to President Bush, and in its latest publications, has strongly endorsed evidence of an average global temperature increase in the 20th century.

There are concerns about possible uncertainties in the instrumental temperature record including the fraction of the globe covered, the effects of changing thermometer designs and observing practices, and the effects of changing land-use around the observing stations. These are being continually addressed in the literature to optimally characterize the accuracy of the record. The casual reader is warned that advocacy organizations considerably exaggerate how seriously these effects contaminate the record. On the other hand considerable skepticism should be exercized for the early years. Indeed, the argument has been advanced that pre-1880 climate models could be more accurate than the instrumental record. Methods standardization and instrument calibration have been the responsibilities of the World Meteorological Organization and its predecessor, the International Meteorological Organization.

There exists a parallel record of marine observations from surface ships. These too suffer from changing practices (such as the switch from collecting water in canvas buckets to measuring the temperature from engine intakes) but they are at least immune to the urban heat island effect. The land and marine records can be compared.

In relation to the instrumental temperature record of the last 100 years, the IPCC Fourth Assessment Report found that: "Urban heat island effects are real but local, and have a negligible influence (less than 0.006 °C per decade over land and zero over the oceans) on these values."

Secondary Data

Secondary evidence for temperature changes can be obtained by observing things that are predicted to be affected by temperature changes, such as variations in the snow cover and ice extent, sea level rise, precipitation, cloud cover, El Niño and extreme weather events. For example, satellite data shows a 10% decrease of snow cover since the late 1960s, and the Northern Hemisphere spring and summer sea-ice extent has decreased by about 10% to 15% since the 1950s and there has been a widespread retreat of mountain glaciers in non-polar regions throughout the 20th century.

Spatial Variability

The global temperature changes are not uniform over the globe, nor would they be expected to be, whether the changes were naturally or humanly forced. Certain places, such as the north shore of Alaska, show dramatic rises in temperature, far above the average for the globe as a whole. The Antarctic peninsula has warmed by 2.5 °C (4.5 °F) in the past five decades in certain places; meanwhile East Antarctic has not significantly warmed.

The Satellite and Balloon Temperature Records

Satellites have been measuring the temperature of the troposphere since 1979. Balloon measurements begin to show an approximation of global coverage in the 1950's.

Several groups have analyzed the satellite data to calculate temperature trends in the troposphere; RSS find a trend of +0.18 °C/decade while UAH find +0.14 °C/decade, to April 2007; Fu et al finds trends (up to the end of 2004) of +0.19 °C/decade when applied to the RSS dataset; Vinnikov and Grody find +0.22°C to +0.26°C per decade (Oct. 03).

Proxies: Tree Rings, Ice Cores: the Last 2000 Years

Longer records exist from proxies: quantities such as tree ring widths, coral growth or isotope variations in ice cores. From these, proxy temperature reconstruction of the last 2000 years have been made for the northern hemisphere. However, coverage of these proxies is sparse: even the best proxy records contain far fewer observations than the worst periods of the observational record. Also, problems exist in connecting the proxies (e.g. tree ring width) to the variable of interest (e.g. temperature).

Indirect Historical Proxies

As well as natural, numerical proxies (tree-ring widths, for example) there exist records from the human historical period which can be used to infer climate variations, often in a less directly numerical way: reports of Frost fairs on the Thames; records of good and bad harvests; dates of spring blossom or lambing; extraordinary falls of rain and snow, and unusual floods or droughts. These too can be used to infer historical temperatures, but generally in a more qualitative manner than the natural proxies discussed above.

Recent evidence suggests that a sudden and short-lived climactic shift between 2200 and 2100 BCE occurred in the region between Tibet and Iceland, with some evidence suggesting a global change. The result was a cooling and reduction in precipitation. This is believed to be a primary cause of the collapse of the Old Kingdom of Egypt.

Paleoclimate

Many estimates of past temperatures have been made over Earth's history. The field of paleoclimatology includes ancient temperature records. As the present article is oriented toward recent temperatures, there is a focus here on events since the retreat of the Pleistocene glaciers. The 10,000 years of the Holocene epoch covers most of this period, since the end of the Northern Hemisphere's Younger Dryas millennium-long cooling. The Holocene Climatic Optimum was generally warmer than the 20th century, but numerous regional variations have been noted since the start of the Younger Dryas.

The Long Term Ice Core Record: The Last 800,000 Years

Even longer term records exist for few sites: the recent Antarctic EPICA core reaches 800 kyr; many others reach more than 100,000 years. The EPICA core covers eight glacial/interglacial cycles. The NGRIP core from Greenland stretchs back 5 kyr into the Eemian interglacial. Whilst the large-scale signals from the cores are clear, there are problems interpreting the detail, and connecting the isotopic variation to the temperature signal.

Geologic Temperature Record

On longer time scales, sediment cores show that the cycles of glacials and interglacials are part of a deepening phase within a prolonged ice age that began with the glaciation of Antarctica approximately 40 million years ago. This deepening phase, and the accompanying cycles, largely began approximately 3 million years ago with the growth of continental ice sheets in the Northern Hemisphere. Gradual changes in Earth's climate of this kind have been frequent during the Earth's 4500 million year existence and most often are attributed to changes in the configuration of continents and ocean sea ways.

The last 3 million years have been characterized by cycles of glacials and interglacials within a gradually deepening ice age. These cycles involve the growth and retreat of continental ice sheets in the Northern Hemisphere and involve fluctuations on 40,000 and 100,000 year time scales. Such cycles are usually interpreted as being driven by predictable changes in the Earth orbit known as Milankovitch cycles. The gradual intensification of this ice age over the last 3 million years has been associated with declining concentrations of the greenhouse gas carbon dioxide, though it remains unclear if this change is sufficiently large to have caused the changes in temperatures.

In the same vein, the initiation of this deepening phase also corresponds roughly to the closure of the Isthmus of Panama by the action of plate tectonics. This prevented direct ocean flow between the Pacific and Atlantic, which would have had significant effects on ocean circulation and the distribution of heat.

However, modeling studies have been ambiguous as to whether this could be the direct cause of the intensification of the present ice age. This recent period of cycling climate is part of the more extended ice age that began ~40 million years ago with the glaciation of Antarctica.

In the earliest part of the Eocene period, a series of abrupt thermal spikes have been observed, lasting no more than a few 100,000 years. The most pronounced of these, the Paleocene-Eocene Thermal Maximum (PETM) is visible in the figure at right. These are usually interpreted as caused by abrupt releases of methane from clathrates (frozen methane ices that accumulate at the bottom of the ocean), though some scientists dispute that methane would be sufficient to cause the observed changes. During these events, temperatures in the Arctic Ocean may have reached levels more typically associated with modern temperate (i.e. mid-latitude) oceans.

During the later portion of the Cretaceous (65-100 million years ago), average global temperatures reached their highest level during the last ~200 million years. This is likely the result of a favorable configuration of the continents during this period that allowed for improved circulation in the oceans and discouraged the formation of large scale ice sheet. Perhaps the visible anecdotal evidence of high temperatures during this period was the occurrence of deciduous forests extending all the way to the poles.

Fluctuations

The Phanerozoic epoch, encompassing the last 542 million years and almost the entire time since the origination of complex multi-celluar life, has more generally been a period of fluctuating temperature between ice ages, such as the current age, and "climate optima", similar to what occurred in the Cretaceous. Roughly 4 such cycles have occurred during this time with an approximately 140 million year separation between climate optima. In addition to the present, ice ages have occurred in the during the Permian-Carboniferous interval and the late Ordovician-early Silurian. There is also a "cooler" interval during the Jurassic and early Cretaceous, with evidence of increased sea ice, but the lack of continents at either pole during this interval prevented the formation of continental ice sheets and consequently this is usually not regarded as a full-fledged ice age. In between these cold period, warmer conditions were present and often referred to as climate optima.

However, it has been difficult to determine whether these warmer intervals were actually hotter or colder than occurred during the Cretaceous optima.

Late Proterozoic Ice Ages

The Neoproterozoic era (542-1000 million years ago), provides evidence of at least two and possibly more major glaciations. The more recent of these ice ages, encompassing the Marinoan & Varangian glacial maxima (~560—~650 million years ago), has been proposed as a snowball Earth event with continuous sea ice reaching nearly to the equator. This is significantly more severe than the ice age during the Phanerozoic. Because this ice age terminated only slightly before the rapid diversification of life during the Cambrian explosion, it has been proposed that this ice age (or at least its end) created conditions favorable to evolution. The earlier Sturtian glacial maxima (~730 million years) may also probably have been a snowball Earth event.

The changes that lead to the initiation of snowball Earth events are not well known, but it has been argued that they necessarily lead to their own end. The widespread sea ice prevents the deposition of fresh carbonates in ocean sediment. Since such carbonates are part of the natural process for recycling carbon dioxide, short-circuiting this process allows carbon dioxide to accumulate in the atmosphere. This increases the greenhouse effect and eventually leads to higher temperatures and the retreat of sea ice.

An Overall View

Direct combination of these interpreted geological temperature records is not necessarily valid, nor is their combination with other more recent temperature records, which may use different definitions. Nevertheless, an overall perspective is useful, even when rather imprecise. Prior to the Neoproterozoic, evidence of temperature changes and glaciation is usually too scattered and sporadic to draw firm conclusions though it seems likely to temperature fluctuations were also substantial during this period. Some evidence does exist however that the period of 2000-3000 million years ago was very generally colder and more glaciated than the last 500 million years. This is usually believed to have resulted from solar radiation approximately 20% lower than today.

On geologically short time scales, greenhouse gases are the most important process influencing temperature. By trapping solar radiation, carbon dioxide, water vapor, methane, and some other gases, contribute to a greenhouse effect essential to maintaining the Earth's temperature. However, modern reconstructions of greenhouse gas concentrations do not show fluctuations over geologically long periods that match to the observed changes in temperatures and ice volumes. This has led most observers to believe that changes in continent configurations are much more important in determining long-term temperature. In particular, the lack of land masses at the poles inhibits the formation of large ice sheets. Similarly, the presence of expansive sea ways, especially near the equator, aids in distributing heat across the globe and leads to higher temperature on average.

According to a recent version, as the solar system moves in and out of the spiral arms of the galaxy, the flux of cosmic rays will vary by approximately a factor of 3. Since such cosmic rays play an important role in ionizing the atmosphere, it is plausible that such variations can influence climate, and may be able to explain the 140 million year quasi-periodicity observed in temperatures during the

Phanerozoic. The biggest present weakness in this theory is that the time period between encounters with galactic spiral arms is only poorly constrained by existing astronomical observations.

On very long time scales, the evolution of the sun is also an important factor in determining Earth's climate. According to standard solar theories, the sun will gradually have increased in brightness as a natural part of its evolution after having started with an intensity approximately 70% of its modern value. The initially low solar radiation, if combined with modern values of greenhouse gases, would not have been sufficient to allow for liquid oceans on the surface of the Earth. However, evidence of liquid water at the surface has been demonstrated till at least 3500 million years ago.

2

GREENHOUSE EFFECT: AN INTRODUCTION

Mistakes are costly and somebody must pay. The time to correct a mistake is before it is made. The cause of mistakes are, first. "I didn't know"; second, "I didn't think"; third, "I did't care."

—*Henry H. Buckley*

What is the Greenhouse Effect?

In common usage, "greenhouse effect" may refer either to the natural greenhouse effect due to naturally occurring greenhouse gases, or to the enhanced (anthropogenic) greenhouse effect which results from gases emitted as a result of human activities.

The greenhouse effect, discovered by Joseph Fourier in 1829 and first investigated quantitatively by Svante Arrhenius in 1896, is the process in which the emission of infrared radiation by the atmosphere warms a planet's surface. The name comes from an analogy with the warming of air inside a greenhouse compared to the air outside the greenhouse. The Earth's average surface temperature is about 20-30°C warmer than it would be without the greenhouse effect. In addition to the Earth, Mars and especially Venus have greenhouse effects.

The Greenhouse Effect is the common term given to the natural build up of greenhouse gases in the lower atmosphere that prevent heat from the sun's rays from escaping back into space. The effect of trapping in the heat is similar to the way a greenhouse works—hence the term 'greenhouse gases' and 'greenhouse effect'. This natural process is crucial for maintaining balanced temperature conditions that are necessary for life on earth.

The Enhanced Greenhouse Effect

However, when the balance of greenhouse gases is disrupted, more heat is trapped and the earth gradually heats up. The growing problem we face today is that human actions are increasing the concentrations of these greenhouse gases well above natural levels, creating the Enhanced Greenhouse Effect, which leads to the associated problems of global warming.

Burning fossil fuels for energy (coal, oil and natural gas) and land clearing further enhance the greenhouse effect, creating the prospect of global climate change. Scientists are concerned, in particular, about rising concentrations of carbon dioxide (CO_2), methane (CH_4), nitrous oxide (N_2O) and various forms of fluorocarbons.

The Greenhouse Effect and the Carbon Cycle

Most fundamentally, life on earth is made possible by energy from the sun, which arrives mainly in the form of visible light. About 30 per cent of sunlight is scattered back into space by the outer atmosphere, but the rest reaches the earth's surface, which reflects it in the form of a calmer, more slow-moving type of energy called infrared radiation. Infrared radiation is carried slowly aloft by air currents, and its eventual escape into space is delayed by greenhouse gases such as water vapour, carbon dioxide, ozone, and methane. Greenhouse gases make up only about 1 per cent of the atmosphere, but they act like a blanket around the earth, or like the glass roof of a greenhouse — they trap heat and keep the planet some 30°C warmer than it would be otherwise. Human activities are making the blanket "thicker" — the natural levels of these gases are being supplemented by emissions of carbon dioxide from the burning of coal, oil, and natural gas; by additional methane and nitrous oxide produced by farming activities and changes in land use; and by several long-lived industrial gases that do not occur naturally.

These changes are happening at unprecedented speed. If emissions continue to grow at current rates, it is almost certain that atmospheric levels of carbon dioxide will double from pre-industrial levels during the 21st century. It is possible they will triple.

The "enhanced greenhouse effect," is a warming of the earth's surface and lower atmosphere. The IPCC assesses with very high confidence that the globally averaged net effect of human activities since 1750 has been one of the factors of warming. The 'best case' computer climate models estimate that the average global temperature will rise by 1.8°C to 4.0°C by the year 2100. A temperature increase of 0.7.4°C occurred last century and for the next two decades, a warming of about 0.2°C per decade is projected should greenhouse gas emissions continue to rise at their current pace and are allowed to double from their pre-industrial level.

As a consequence, rise in temperature will be accompanied by changes in climate — in such things as cloud cover, precipitation, wind patterns, and the duration of seasons. As per its Fourth Assessment Report, the IPCC projects that heat waves and heavy precipitation events are very likely to increase in frequency in the 21st century. In a world that is crowded and under stress, millions of people depend on weather patterns, such as monsoon rains, to continue as they have in the past. Changes, at a minimum, will be difficult and disruptive.

As is evident, Carbon dioxide is responsible for over 60 per cent of the "enhanced greenhouse effect." Which is produced by are burning coal, oil, and natural gas at a rate that is much, much faster than the speed at which these fossil fuels were created. This is releasing the carbon stored in the fuels into the atmosphere and upsetting the carbon cycle, the millennia-old, precisely balanced system by which carbon is exchanged between the air, the oceans, and land vegetation. As of now atmospheric levels of carbon dioxide are rising by over 10 per cent every 20 years. Fossil fuel use is boosting atmospheric levels of carbon, upsetting an age-old balance.

Notably, climate change is inevitable because of past and current emissions. The climate does not respond immediately to external changes, but after 150 years of industrialization, global warming has momentum, and it will continue to affect the earth's natural systems for hundreds of years even if greenhouse gas emissions are reduced and atmospheric levels stop rising.

Greenhouse Gases help us live comfortably in Earth. Greenhouses gases trap the Sun's warmth by not allowing natural gases to escape. The problem is that humans are burning more fossil fuels than ever before and are cutting down thousands of trees causing more and more greenhouse gases. The more gases that there are, the more the earth heats up. We are adding more to the natural greenhouse gases, and as result the earth is getting hotter and the ice in Antarctica is melting. The climate is changing and is causing harm in animals and plants habitats. As a natural phenomenon usually the average temperatures on earth increases and decreases, but in the past decades the earth's temperatures are rising. Some scientist believe that this is happening because we are using more gas, oil, coal, methane.

The greenhouse effect has increased since the Industrial Revolution. Humans are burning more fossil fuels and are using more gases like carbon dioxide, methane and nitrous oxide more than ever. These activities are causing the earth to warm up and, as a result, global warming is happening.

Most researchers agree that even small temperature changes can send hundreds of already struggling species into extinction unless we stem the tide of global warming.

A report by a United Nations group of international scientists makes it clear that the consequences of global warming are no longer avoidable and that human activity has been the primary cause of rapid global warming for more than half a century.

Scientists have long predicted and recently measured notable increases in the world's temperature due to the Enhanced Greenhouse Effect. Since the 1990's average temperature have increased 0.7°C. The ten warmest years on record have all occurred since 1987 with nine of them since 1990. The effect of climate change will vary around the globe. Greater temperature rises are expected towards the poles than near the Equator, and more significant over land than at sea. It is possible that there will be both positive and negative impacts. These impacts are predicted to affect the:

(i) distribution of plants and animals;

(ii) sea levels;

(iii) frequency of storms and floods; and

(iv) spread of weeds, pests and diseases, possibly influencing agriculture.

The Basic Mechanism

A comprehensive way to illustrate the basic mechanism of Greenhouse effect is presented below. The Earth receives energy from the Sun in the form of radiation. The Earth reflects about 30% of the incident solar flux; the remaining 70% is absorbed, warming the land, atmosphere and oceans.

To the extent that the Earth is in a steady state, the energy stored in the atmosphere and ocean does not change in time, so energy equal to the absorbed solar radiation must be radiated back to space. Earth radiates energy into space as black-body radiation, which maintains a thermal equilibrium. This thermal, infrared radiation increases with increasing temperature. One can think of the Earth's temperature as being determined by the infrared flux needed to balance the absorbed solar flux.

The visible solar radiation heats the surface, not the atmosphere, whereas most of the infrared radiation escaping to space is emitted from the upper atmosphere, not the surface. The infrared photons emitted by the surface are mostly absorbed by the atmosphere and do not escape directly to space. The reason this warms the surface is most easily understood by starting with a simplified model of a purely radiative greenhouse effect that ignores energy transfer in the atmosphere by convection (sensible heat transport) and by the evaporation and condensation of water vapor (latent heat transport). In this purely radiative case, one can think of the atmosphere as emitting infrared radiation both upwards and downwards. The upward infrared flux emitted by the surface must balance not only the absorbed solar flux but also this downward infrared flux emitted by the atmosphere.

The surface temperature will rise until it generates thermal radiation equivalent to the sum of these two incident radiation streams. A more realistic picture taking into account the convective and latent heat fluxes is somewhat more complex. But the following simple model captures the essence. The starting point is to note that the opacity of the atmosphere to infrared radiation determines the height in the atmosphere from which most of the photons emitted to space are emitted. If the atmosphere is more opaque, the typical photon escaping to space will be emitted from higher in the atmosphere, because one then has to go to higher altitudes to *see* out to space in the infrared. Since the emission of infrared radiation is a function of temperature, it is the temperature of the atmosphere at this emission level that is effectively determined by the requirement that the emitted flux balance the absorbed solar flux. But the temperature of the atmosphere generally decreases with height above the surface, at a rate of roughly 6.5 °C per kilometer on average, until one reaches the stratosphere 10-15 km above the surface. (Most infrared photons escaping to space are emitted by the troposphere, the region bounded by the surface and the stratosphere, so we can ignore the stratosphere in this simple picture.) A very simple model, but one that proves to be remarkably useful, involves the assumption that this temperature profile is simply fixed, by the non-radiative energy fluxes. Given the temperature at the emission level of the infrared flux escaping to space, one then computes the surface temperature by increasing temperature at the rate of 6.5 °C per kilometer, the environmental lapse rate, until one reaches the surface. The more opaque the atmosphere, and the higher the emission level of the escaping infrared radiation, the warmer the surface, since one then needs to follow this lapse rate over a larger distance in the vertical.

The Greenhouse Gases

Quantum mechanics provides the basis for computing the interactions between molecules and radiation. Most of this interaction occurs when the frequency of the radiation closely matches that of the spectral lines of the molecule, determined by the quantization of the modes of vibration and rotation of the molecule. (The electronic excitations are generally not relevant for infrared radiation, as they require energy larger than that in an infrared photon.)

The width of a spectral line is an important element in understanding its importance for the absorption of radiation. In the Earth's atmosphere these spectral widths are primarily determined by "pressure broadening", which is the distortion of the spectrum due to the collision with another molecule. Most of the infrared absorption in the atmosphere can be thought of as occurring while two molecules are colliding. The absorption due to a photon interacting with a lone molecule is relatively small. This three-body aspect of the problem, one photon and two molecules, makes direct quantum mechanical computation for molecules of interest more challenging. Careful laboratory spectroscopic measurements, rather than *ab initio* quantum mechanical computations, provide the basis for most of the radiative transfer calculations used in studies of the atmosphere.

The molecules/atoms that constitute the bulk of the atmosphere; oxygen (O_2), nitrogen (N_2) and argon; do not interact with infrared radiation significantly. While the oxygen and nitrogen molecules can vibrate, because of their symmetry these vibrations do not create any transient charge separation that enhances the interaction with radiation. In the Earth's atmosphere, the dominant infrared absorbing gases are water vapor, carbon dioxide, and ozone (O_3), these molecules being "floppier" so that their rotation/vibration modes are more easily excited. For example, carbon dioxide is a linear molecule, but it has an important vibrational mode in which the molecule bends with the carbon in the middle moving one way and the oxygens on the ends moving the other way, creating some charge separation, a dipole moment. A substantial part of the greenhouse effect due to carbon dioxide exists because this vibration is easily excited by infrared radiation. Clouds are also very important infrared absorbers. Therefore, water has multiple effects on infrared radiation, through its vapor phase and through its condensed phases. Other absorbers of significance include methane, nitrous oxide and the chlorofluorocarbons.

Discussion of the relative importance of different infrared absorbers is confused by the overlap between the spectral lines due to different gases, widened by pressure broadening. As a result, the absorption due to one gas cannot be thought of as independent of the presence of other gases. One convenient approach is to remove the chosen constituent, leaving all other absorbers, and the temperatures, untouched, and monitoring the infrared radiation escaping to space. The reduction in infrared absorbtion is then a measure of the importance of that constituent. More precisely, define the greenhouse effect (GE) to be the difference between the infrared radiation that the surface would radiate to space if there were no atmosphere and the actual infrared radiation escaping to space. Then compute the percentage reduction in GE when a constituent is removed. The table below is computed by this method, using a particular 1-dimensional model of the atmosphere. More recent 3D computations lead to similar results.

Gas Removed	Percent Reduction in GE
H_2O	36%
CO_2	12%
O_3	3%

(*Source:* Ramanathan and Coakley, *Rev. Geophys and Space Phys.,* 16 465 (1978)).

By this particular measure, water vapor can be thought of as providing 36% of the greenhouse effect, and carbon dioxide 12%, but the effect of removal of both of these constituents will be greater than 48%. An additional proviso is that these numbers are computed holding the cloud distribution fixed. But removing water vapor from the atmosphere while holding clouds fixed is not likely to be physically relevant. In addition, the effects of a given gas are typically nonlinear in the amount of that gas, since the absorption by the gas at one level in the atmosphere can remove photons that would otherwise interact with the gas at another altitude. The kinds of estimates presented in the table, while often encountered in the controversies surrounding global warming, must be treated with caution. Different estimates found in different sources typically result from different definitions and do not reflect uncertainties in the underlying radiative transfer.

Positive Feedback and Runaway Greenhouse Effect

When the concentration of a greenhouse gas (A) is itself a function of temperature, there is a positive feedback from the increase in another greenhouse gas (B), whereby increase in B increases the temperature which, in turn, increases the concentration of A, which increases temperatures further, and so on. This feedback is bound to stop, since the overall supply of the gas A must be finite. If this feedback ends after producing a major temperature increase, it is called a runaway greenhouse effect.

According to some climate models (Clathrate gun hypothesis), such a runaway greenhouse effect, involving liberation of methane gas from hydrates by global warming, caused the Permian-Triassic extinction event. It is also thought that large quantities of methane could be released from the Siberian tundra as it begins to thaw, methane being 21-times more potent a greenhouse gas than carbon dioxide.

A runaway greenhouse effect involving CO_2 and water vapor may have occurred on Venus. On Venus today there is little water vapor in the atmosphere. If water vapor did contribute to the warmth of Venus at one time, this water is thought to have escaped to space. Venus is sufficiently strongly heated by the Sun that water vapor can rise much higher in the atmosphere and is split into hydrogen and oxygen by ultraviolet light. The hydrogen can then escape from the atmosphere and the oxygen recombines. Carbon dioxide, the dominant greenhouse gas in the current Venusian atmosphere, likely owes its larger concentration to the weakness of carbon recycling as compared to Earth, where the carbon dioxide emitted from volcanoes is efficiently subducted into the Earth by plate tectonics on geologic time scales.

Anthropogenic Greenhouse Effect

CO_2 production from increased industrial activity (fossil fuel burning) and other human activities such as cement production and tropical deforestation has increased the

CO_2 concentrations in the atmosphere. Measurements of carbon dioxide amounts from Mauna Loa observatory show that CO_2 has increased from about 313 ppm (parts per million) in 1960 to about 375 ppm in 2005. The current observed amount of CO_2 exceeds the geological record of CO_2 maxima (~300 ppm) from ice core data (Hansen, J., Climatic Change, 68, 269, 2005).

Because it is a greenhouse gas, elevated CO_2 levels will increase global mean temperature; based on an extensive review of the scientific literature, the Intergovernmental Panel on Climate Change concludes that "*most of the observed increase in globally averaged temperatures since the mid-20th century is very likely due to the observed increase in anthropogenic greenhouse gas concentrations*".

Over the past 800,000 years, ice core data shows unambiguously that carbon dixoide has varied from values as low as 180 parts per million (ppm) to the pre-industrial level of 270ppm. Certain paleoclimatologists consider variations in carbon dioxide to be a fundamental factor in controlling climate variations over this time scale.

The term 'greenhouse effect' originally came from the greenhouses used for gardening, but it is a misnomer since greenhouses operate differently. A greenhouse is built of glass; it heats up primarily because the Sun warms the ground inside it, which warms the air near the ground, and this air is prevented from rising and flowing away. The warming inside a greenhouse thus occurs by suppressing convection and turbulent mixing. This can be demonstrated by opening a small window near the roof of a greenhouse: the temperature will drop considerably. It has also been demonstrated experimentally (Wood, 1909): a "greenhouse" built of rock salt (which is transparent to IR) heats up just as one built of glass does. Greenhouses thus work primarily by preventing convection; the atmospheric greenhouse effect however reduces radiation loss, not convection. It is quite common, however, to find sources that make the "greenhouse" analogy. Although the primary mechanism for warming greenhouses is the prevention of mixing with the free atmosphere, the radiative properties of the glazing can still be important to commercial growers. With the modern development of new plastic surfaces and glazings for greenhouses, this has permitted construction of greenhouses which selectively control radiation transmittance in order to better control the growing environment.

When sunlight reaches the surface of the Earth, some of it is absorbed and warms the Earth. Because the Earth's surface is much cooler than the sun, it radiates energy at much longer wavelengths than does the sun. The atmosphere absorbs these longer wavelengths more effectively than it does the shorter wavelengths from the sun. The absorption of this longwave radiant energy warms the atmosphere; the atmosphere also is warmed by transfer of sensible and latent heat from the surface. Greenhouse gases also *emit* longwave radiation both upward to space and downward to the surface. The downward part of this longwave radiation emitted by the atmosphere is the "greenhouse effect." The term is a misnomer, as this process is not the mechanism that warms greenhouses.

The major natural greenhouse gases are water vapor, which causes about 36-70% of the greenhouse effect on Earth (not including clouds); carbon dioxide, which causes 9-26%; methane, which causes 4-9%, and ozone, which causes 3-7%. It is not possible to state that a certain gas causes a certain percentage of the greenhouse effect, because the influences of the various gases are not additive. (The higher ends of the ranges quoted are for the gas alone; the lower ends, for the gas counting overlaps.) Other greenhouse gases include, but are not limited to, nitrous oxide, sulfur hexafluoride, hydrofluorocarbons, perfluorocarbons and chlorofluorocarbons.

It must be remembered that the major atmospheric constituents (nitrogen, N_2 and oxygen, O_2) are not greenhouse gases. This is because homonuclear diatomic molecules such as N_2 and O_2 neither absorb nor emit infrared radiation, as there is no net change in the dipole moment of these molecules when they vibrate. Molecular vibrations occur at energies that are of the same magnitude as the energy of the photons on infrared light.

It is worth noting that late 19th century scientists experimentally discovered that N_2 and O_2 did not absorb infrared radiation (called, at that time, "dark radiation") and that CO_2 and many other gases did absorb such radiation. It was recognized in the early 20th century that the known major greenhouse gases in the atmosphere did cause the earth's temperature to be higher than it would have been without the greenhouse gases.

Anthropogenic Greenhouse Gases

The concentrations of several greenhouse gases have increased over time. Human activity increases the greenhouse effect primarily through release of carbon dioxide, but human influences on other greenhouse gases can also be important. Some of the main sources of greenhouse gases due to human activity include:

- burning of fossil fuels and deforestation leading to higher carbon dioxide concentrations;

- livestock and paddy rice farming, land use and wetland changes, pipeline losses, and covered vented landfill emissions leading to higher methane atmospheric concentrations. Many of the newer style fully vented septic systems that enhance and target the fermentation process also are major sources of atmospheric methane;
- use of chlorofluorocarbons (CFCs) in refrigeration systems, and use of CFCs and halons in fire suppression systems and manufacturing processes.
- agricultural activities, including the use of fertilizers, that lead to higher nitrous oxide concentrations.

Greenhouse gas emissions from industry, transportation and agriculture are very likely the main cause of recently observed global warming.

Carbon dioxide, methane, nitrous oxide and three groups of fluorinated gasses (sulfur hexafluoride, HFCs, and PFCs) are the major greenhouse gases and the subject of the Kyoto Protocol, which entered into force in 2005.

CFCs, although greenhouse gases, are regulated by the Montreal Protocol, which was motivated by CFCs' contribution to ozone depletion rather than by their contribution to global warming. Note that ozone depletion has only a minor role in greenhouse warming though the two processes often are confused in the popular media.

The Role of Water Vapor

Water vapor is a naturally occurring greenhouse gas and accounts for the largest percentage of the greenhouse effect, between 36% and 90%. Water vapor concentrations fluctuate regionally, but human activity does not directly affect water vapor concentrations except at local scales.

In climate models an increase in atmospheric temperature caused by the greenhouse effect due to anthropogenic gases will in turn lead to an increase in the water vapor content of the troposphere, with approximately constant relative humidity. The increased water vapor in turn leads to an increase in the greenhouse effect and thus a further increase in temperature; the increase in temperature leads to still further increase in atmospheric water vapor; and the feedback cycle continues until equilibrium is reached. Thus water vapor acts as a positive feedback to the forcing provided by human-released greenhouse gases such as CO_2.

Increase of Greenhouse Gases

Measurements from Antarctic ice cores show that just before industrial emissions began, atmospheric CO_2 levels were about 280 parts per million by volume (ppm; the units $\mu L/L$ are occasionally used and are identical to parts per million by volume). From the same ice cores it appears that CO_2 concentrations stayed between 260 and 280 ppm during the preceding 10,000 years. Studies using evidence from stomata of fossilized leaves suggest greater variability, with CO_2 levels above 300 ppm during the period 7,000-10,000 years ago, though others have argued that these findings more likely reflect calibration/contamination problems rather than actual CO_2 variability.

Since the beginning of the Industrial Revolution, the concentrations of many of the greenhouse gases have increased. The concentration of CO_2 has increased by about 100 ppm (i.e., from 280 ppm to 380 ppm). The first 50 ppm increase took place in about 200 years, from the start of the Industrial Revolution to around 1973; the next 50 ppm increase took place in about 33 years, from 1973 to 2006. Many observations are available on line in a variety of Atmospheric Chemistry Observational Databases.

Removal from the Atmosphere and Global Warming Potential

Aside from water vapor near the surface, which has a residence time of days, most greenhouse gases take a very long time to leave the atmosphere. Although it is not easy to know with precision how long, there are estimates of the duration of stay, i.e., the time which is necessary so that the gas disappears from the atmosphere, for the principal greenhouse gases. Greenhouse gases can be removed from the atmosphere by various processes:

- as a consequence of a physical change (condensation and precipitation remove water vapor from the atmosphere).
- as a consequence of chemical reactions within the atmosphere. This is the case for methane. It is oxidized by reaction with naturally occurring hydroxyl radical, OH· and degraded to CO_2 and water vapor at the end of a chain of reactions (the contribution of the CO_2 from the oxidation of methane is not included in the methane Global warming potential). This also includes solution and solid phase chemistry occurring in atmospheric aerosols.
- as a consequence of a physical interchange at the interface between the atmosphere and the other compartments of the planet. An example is the mixing of atmospheric gases into the oceans at the boundary layer.

- as a consequence of a chemical change at the interface between the atmosphere and the other compartments of the planet. This is the case for CO_2, which is reduced by photosynthesis of plants, and which, after dissolving in the oceans, reacts to form carbonic acid and bicarbonate and carbonate ions.
- as a consequence of a photochemical change. Halocarbons are dissociated by UV light releasing Cl· and F· as free radicals in the stratosphere with harmful effects on ozone (halocarbons are generally too stable to disappear by chemical reaction in the atmosphere).
- as a consequence of dissociative ionization caused by high energy cosmic rays or lightning discharges, which break molecular bonds. For example, lightning forms N anions from N_2 which then react with O_2 to form NO_2.

Two scales can be used to describe the effect of different gases in the atmosphere. The first, the atmospheric lifetime, describes how long it takes to restore the system to equilibrium following a small increase in the concentration of the gas in the atmosphere. Individual molecules may interchange with other reservoirs such as soil, the oceans, and biological systems, but the mean lifetime refers to the decaying away of the excess. It is sometimes erroneously claimed that the atmospheric lifetime of CO_2 is only a few years because that is the average time for any CO_2 molecule to stay in the atmosphere before being removed by mixing into the ocean, uptake by photosynthesis, or other processes. This ignores the balancing fluxes of CO_2 into the atmosphere from the other reservoirs. It is the net concentration changes of the various greenhouse gases by *all sources and sinks* that determines atmospheric lifetime, not just the removal processes.

The second scale is global warming potential (GWP). The GWP depends on both the efficiency of the molecule as a greenhouse gas and its atmospheric lifetime. GWP is measured relative to the same mass of CO_2 and evaluated for a specific timescale. Thus, if a molecule has a high GWP on a short time scale (say 20 years) but has only a short lifetime, it will have a large GWP on a 20 year scale but a small one on a 100 year scale. Conversely, if a molecule has a longer atmospheric lifetime than CO_2 its GWP will increase with time.

Examples of the atmospheric lifetime and GWP for several greenhouse gases include:

- CO_2 has a variable atmospheric lifetime (approximately 200-450 years for small perturbations). Recent work indicates that recovery from a large input of atmospheric CO_2 from burning fossil fuels will result in an effective lifetime of tens of thousands of years. Carbon dioxide is defined to have a GWP of 1 over all time periods.
- Methane has an atmospheric lifetime of 12 ± 3 years and a GWP of 62 over 20 years, 23 over 100 years and 7 over 500 years. The decrease in GWP associated with longer times is associated with the fact that the methane is degraded to water and CO_2 by chemical reactions in the atmosphere.
- Nitrous oxide has an atmospheric lifetime of 120 years and a GWP of 296 over 100 years.

- CFC-12 has an atmospheric lifetime of 100 years and a GWP(100) of 10600.
- HCFC-22 has an atmospheric lifetime of 12.1 years and a GWP(100) of 1700.
- Tetrafluoromethane has an atmospheric lifetime of 50,000 years and a GWP(100) of 5700.
- Sulfur hexafluoride has an atmospheric lifetime of 3,200 years and a GWP(100) of 22000.

Source : IPCC, table 6.7.

Carbon monoxide has an indirect radiative effect by elevating concentrations of methane and tropospheric ozone through scavenging of atmospheric constituents (e.g., the hydroxyl radical, OH) that would otherwise destroy them. Carbon monoxide is created when carbon-containing fuels are burned incompletely. Through natural processes in the atmosphere, it is eventually oxidized to carbon dioxide. Carbon monoxide has an atmospheric lifetime of only a few months and as a consequence is spatially more variable than longer-lived gases.

Another potentially important indirect effect comes from methane, which in addition to its direct radiative impact also contributes to ozone formation. Shindell et al (2005) argue that the contribution to climate change from methane is at least double previous estimates as a result of this effect.

REFERENCES

Bowen, Mark; Thin Ice: Unlocking the Secrets of Climate in the World's Highest Mountains; Owl Books, 2005.

Earth Radiation Budget, http://marine.rutgers.edu/mrs/education/class/yuri/erb.html

Fleagle, RG and Businger, JA: An introduction to atmospheric physics, 2nd edition, 1980.

Fraser, Alistair B., Bad Greenhouse http://www.ems.psu.edu/~fraser/Bad/BadGreenhouse.html

Giacomelli, Gene A. and William J. Roberts1, Greenhouse Covering Systems, Rutgers University, downloaded from: http://ag.arizona.edu/ceac/research/archive/HortGlazing.

Ann Henderson-Sellers and McGuffie, K: A climate modelling primer.

Idso, S.B.: Carbon Dioxide: friend or foe, 1982.

Kiehl, J.T., and Trenberth, K. (1997). Earth's annual mean global energy budget, *Bulletin of the American Meteorological Society* 78 (2), 197–208.

Piexoto, JP and Oort, AH: Physics of Climate, American Institute of Physics, 1992.

Wood, R.W. (1909). Note on the Theory of the Greenhouse, *Philosophical Magazine* 17, p319–320. For the text of this online, see http://www.wmconnolley.org.uk/sci/wood_rw.1909.html

IPCC assessment reports, see http://www.ipcc.ch/

How to kill (almost) all life: the end-Permian extinction event, Michael J. Benton and Richard J. Twitchett, Department of Earth Sciences University of Bristol UK, Trends in Ecology and Evolution, Vol.18, No.7 July 2003.

3

CLIMATE CHANGE: WHAT & WHY?

> There is nothing more permanent than change, and nothing that meets with more resistance. Yet nothing stands still. Everything either moves forward or falls backward. Man has progressed through the centuries because man has had the intelligence to meet change with change. Individuals, too, must heed that lesson of the race. And, of course, men can master change by training today for the opportunities which new conditions will open tomorrow. It was good advice a generation or so ago, and it is even better advice today.

There is no denying the fact that climate change is happening, and it is linked directly to human activities that emit greenhouse gases. As explored, a number of technological options exist to avert dangerous climatic change by dramatically reducing greenhouse gas emissions both now and into the future. To successfully address the enormous challenge of climate change, new approaches are needed at the international level, and the United States must re-engage in the global effort and adopt strong and emphatic national policies.

Climate change refers to the variation in the Earth's global climate or in regional climates over time. It describes changes in the variability or average state of the atmosphere over time scales ranging from decades to millions of years.

These changes can be caused by processes internal to the Earth, external forces (e.g. variations in sunlight intensity) or, more recently, human activities.

In recent usage, especially in the context of environmental policy, the term "climate change" often refers only to changes in modern climate, including the rise in average surface temperature known as global warming. In some cases, the term is also used with a presumption of human causation, as in the United Nations Framework Convention on Climate Change (UNFCCC). The UNFCCC uses "climate variability" for non-human caused variations.

Major Factors Responsible for Climate Change

Climate changes reflect variations within the Earth's atmosphere, processes in other parts of the Earth such as oceans and ice caps, and the impact of human activity. The external factors that can shape climate are often called climate forcings and include such processes as variations in solar radiation, the Earth's orbit, and greenhouse gas concentrations.

Variations Within the Earth's Climate

Weather is the day-to-day state of the atmosphere, and is a chaotic non-linear dynamical system. On the other hand, *climate* — the average state of weather — is fairly stable and predictable. Climate includes the average temperature, amount of precipitation, days of sunlight, and other variables that might be measured at any given site. However, there are also changes within the Earth's environment that can affect the climate. According to NASA the average global temperature is currently 14.6°C.

Glaciation: Glaciers are recognized as one of the most sensitive indicators of climate change, advancing substantially during climate cooling (e.g., the Little Ice Age) and retreating during climate warming on moderate time scales. Glaciers grow and collapse, both contributing to natural variability and greatly amplifying externally-forced changes. For the last century, however, glaciers have been unable to regenerate enough ice during the winters to make up for the ice lost during the summer months.

The most significant climate processes of the last several million years are the glacial and interglacial cycles of the present ice age. Though shaped by orbital variations, the internal responses involving continental ice sheets and 130 m sea-level change certainly played a key role in deciding what climate response would be observed in most regions. Other changes, including Heinrich events, Dansgaard–Oeschger events and the Younger Dryas show the potential for glacial variations to influence climate even in the absence of specific orbital changes.

Ocean Variability: On the decadal scale of decades, climate changes can also result from interaction of the atmosphere and oceans. Many climate fluctuations, the best known being the El Niño Southern oscillation but also including the Pacific decadal oscillation, the North Atlantic oscillation, and the Arctic oscillation, owe their existence at least in part to different ways that heat can be stored in the oceans and move between different reservoirs. On longer time scales ocean processes such as thermohaline circulation play a key role in redistributing heat, and can dramatically affect climate.

Generally, most forms of internal variability in the climate system can be recognized as a form of hysteresis, meaning that the current state of climate reflects not only the inputs, but also the history of how it got there. For example, a decade of dry conditions may cause lakes to shrink, plains to dry up and deserts to expand. In turn, these conditions may lead to less rainfall in the following years. In short, climate change can be a self-perpetuating process because different aspects of the environment respond at different rates and in different ways to the fluctuations that inevitably occur.

Non-Climate Factors Driving Climate Change

Greenhouse Gases: In fact, radiative forcing by greenhouse gases is the primary cause of global warming. Greenhouse gases are also important in understanding Earth's climate history. According to these studies, the greenhouse effect, which is the warming produced as greenhouse gases trap heat, plays a key role in regulating Earth's temperature. Over the last 600 million years, carbon dioxide concentrations have varied from perhaps >5000 ppm to less than 200 ppm, due primarily to the impact of geological processes and biological innovations. It has been argued by Veizer et al., 1999, that variations in greenhouse gas concentrations over tens of millions of years have not been well correlated to climate change, with plate tectonics perhaps playing a more dominant role. More recently Royer et al. have used the CO_2-climate correlation to derive a value for the climate sensitivity. There are several examples of rapid changes in the concentrations of greenhouse gases in the Earth's atmosphere that do appear to correlate to strong warming, including the Paleocene–Eocene thermal maximum, the Permian–Triassic extinction event, and the end of the Varangian snowball earth event.

Presently, , rising carbon dioxide levels are implicated as the primary cause of global warming since 1950. According to the Intergovernmental Panel on Climate Change (IPCC),

2007, the atmospheric concentration of CO_2 in 2005 was 379ppm[3] compared to the pre-industrial levels of 280ppm[3].

Thermodynamics and Le Chatelier's principle explain the characteristics of the dynamic equilibrium of a gas in solution such as the vast amount of $C0_2$ held in solution in the world's oceans moving into and returning from the atmosphere. These principals can be observed as bubbles which rise in a pot of water heated on a stove, or a in a glass of cold beer allowed to sit at room temperature; gases dissolved in liquids are released under certain circumstances.

Plate Tectonics: This is also a natural phenomenon on the longest time scales, plate tectonics will reposition continents, shape oceans, build and tear down mountains and generally serve to define the stage upon which climate exists. More recently, plate motions have been implicated in the intensification of the present ice age when, approximately 3 million years ago, the North and South American plates collided to form the Isthmus of Panama and shut off direct mixing between the Atlantic and Pacific Oceans.

Solar Variation: Basically, the sun is the ultimate source of essentially all heat in the climate system. The energy output of the sun, which is converted to heat at the Earth's surface, is an integral part of shaping the Earth's climate. On the longest time scales, the sun itself is getting brighter with higher energy output; as it continues its main sequence, this slow change or evolution affects the Earth's atmosphere. Early in Earth's history, it is thought to have been too cold to support liquid water at the Earth's surface, leading to what is known as the Faint young sun paradox. On more modern time scales, there are also a variety of forms of solar variation, including the 11-year solar cycle and longer-term modulations.

However, the 11-year sunspot cycle does not manifest itself clearly in the climatological data. Solar intensity variations are considered to have been influential in triggering the Little Ice Age, and for some of the warming observed from 1900 to 1950. The cyclical nature of the sun's energy output is not yet fully understood; it differs from the very slow change that is occurring to the sun as it ages and evolves.

Orbital Variations: In their impact on climate, orbital variations are in some sense an extension of solar variability, because slight variations in the Earth's orbit lead to changes in the distribution and abundance of sunlight reaching the Earth's surface. Such orbital variations, known as Milankovitch cycles, are a highly predictable consequence of basic physics due to the mutual interactions of the Earth, its moon, and the other planets. These variations are considered the driving factors underlying the glacial and interglacial cycles of the present ice age. Subtler variations are also present, such as the repeated advance and retreat of the Sahara desert in response to orbital precession.

Volcanism: A single eruption of the kind that occurs several times per century can impact climate, causing cooling for a period of a few years. For example, the eruption of Mount Pinatubo in 1991 is barely visible on the global temperature profile. Huge eruptions, known as large igneous provinces, occur only a few times every hundred million years, but can reshape climate for millions of years and cause mass extinctions. Initially, scientists thought that the dust emitted into the atmosphere from large volcanic eruptions was responsible for the cooling by partially blocking the transmission of solar radiation to the Earth's surface. However, measurements indicate that most of the dust thrown in the atmosphere returns to the Earth's surface within six months.

Volcanos are also part of the extended carbon cycle. Over very long (geological) time periods, they release carbon dioxide from the earth's interior, counteracting the uptake by sedimentary rocks and other geological carbon sinks. However, this contribution is insignificant compared to the current anthropogenic emissons. The US Geological Survey estimates that human activities generate 150 times the amount of carbon dioxide emitted by volcanoes.

Human Influences on Climate Change

The biggest factor of present concern is the increase in CO_2 levels due to emissions from fossil fuel combustion, followed by aerosols (particulate matter in the atmosphere) which exerts a cooling effect. Other factors, including land use, ozone depletion, animal agriculture and deforestation also impact climate.

Fossil Fuels: Beginning with the industrial revolution in the 1850s and accelerating ever since, the human consumption of fossil fuels has elevated CO_2 levels from a concentration of ~280 ppm to more than 380 ppm today. These increases are projected to reach more than 560 ppm before the end of the 21st century. It is known that carbon dioxide levels are substantially higher now than at any time in the last 800,000 years. Along with rising methane levels, these changes are anticipated to cause an increase of 1.4–5.6 °C between 1990 and 2100.

Anthropogenic aerosols, particularly sulphate aerosols from fossil fuel combustion, are believed to exert a cooling influence. This, together with natural variability, is believed to account for the relative "plateau" in the graph of 20th century temperatures in the middle of the century.

Irrigation, deforestation, and agriculture fundamentally change the environment. For example, they change the amount of water going into and out of a given location. They also may change the local albedo by influencing the ground cover and altering the amount of sunlight that is absorbed. For example, there is evidence to suggest that the climate of Greece and other Mediterranean countries was permanently changed by widespread deforestation between 700 BC and 1 AD (the wood being used for shipbuilding, construction and fuel), with the result that the modern climate in the region is significantly hotter and drier, and the species of trees that were used for shipbuilding in the ancient world can no longer be found in the area.

A controversial hypothesis by William Ruddiman called the early anthropocene hypothesis suggests that the rise of agriculture and the accompanying deforestation led to the increases in carbon dioxide and methane during the period 5000–8000 years ago. These increases, which reversed previous declines, may have been responsible for delaying the onset of the next glacial period, according to Ruddimann's overdue-glaciation hypothesis.

In modern times, a 2007 Jet Propulsion Laboratory study found that the average temperature of California has risen about 2 degrees over the past 50 years, with a much higher increase in urban areas. The change was attributed mostly to extensive human development of the landscape.

Livestock: It surprising to note that according to a 2006 United Nations report, livestock is responsible for 18% of the world's greenhouse gas emissions as measured in CO_2 equivalents. This however includes land usage change, meaning deforestation in order to create grazing land. In the Amazon, 70% of deforestation is to make way for grazing land, so this is the major factor in the 2006 UN FAO report, which was the first agricultural report to include land usage change into the radiative forcing of livestock. In addition to CO_2 emissions, livestock produces 65% of human-induced nitrous oxide (which has 296 times the global warming potential of CO_2) and 37% of human-induced methane (which has 23 times the global warming potential of CO_2).

If a certain forcing (for example, solar variation) acts to change the climate, then there may be mechanisms that act to amplify or reduce the effects. These are called positive and negative feedbacks. As far as is known, the climate system is generally stable with respect to these feedbacks: positive feedbacks do not "run away". Part of the reason for this is the existence of a powerful negative feedback between temperature and emitted radiation: radiation increases as the fourth power of absolute temperature.

However, a number of important positive feedbacks do exist. The glacial and interglacial cycles of the present ice age provide an important example. It is believed that orbital variations provide the timing for the growth and retreat of ice sheets. However, the ice sheets themselves reflect sunlight back into space and hence promote cooling and their own growth, known as the ice-albedo feedback. Further, falling sea levels and expanding ice decrease plant growth and indirectly lead to declines in carbon dioxide and methane. This leads to further cooling. Similarly, rising temperatures caused, for example, by anthropogenic emissions of greenhouse gases could lead to retreating snow lines, revealing darker ground underneath, and consequently result in more absorption of sunlight.

Water vapor, methane, and carbon dioxide can also act as significant positive feedbacks, their levels rising in response to a warming trend, thereby accelerating that trend. Water vapor acts strictly as a feedback (excepting small amounts in the stratosphere), unlike the other major greenhouse gases, which can also act as forcings.

More complex feedbacks involve the possibility of changing circulation patterns in the ocean or atmosphere. For example, a significant concern in the modern case is that melting glacial ice from Greenland will interfere with sinking waters in the North Atlantic and inhibit thermohaline circulation. This could affect the Gulf Stream and the distribution of heat to Europe and the east coast of the United States.

Other potential feedbacks are not well understood and may either inhibit or promote warming. For example, it is unclear whether rising temperatures promote or inhibit vegetative growth, which could in turn draw down either more or less carbon dioxide. Similarly, increasing temperatures may lead to either more or less cloud cover. Since on balance cloud cover has a strong cooling effect, any change to the abundance of clouds also impacts climate.

Humanitarian Impact

The events of climate-change are extreme, in terms of intensity, which often cause extensive damage and substantial loss of life in a developing country. They are repeated, and would therefore have an economic and social impact difficult to afford for poor countries. They are difficult to foresee because they are not following the "traditional" climate cycle.

1. The events of climate-change are a threat to

human life –Events of high intensity would have an impact in terms of human loss. The rise of the sea level will also endanger coastal cities, such as Lagos.

2. They are a threat to food security and water supply. Climate change will increasingly put pressure on crop yields. Indeed, the decrease of agricultural productivity in tropical and sub-tropical zones particularly will endanger food security. As well as the warming of oceans, which will affect communities, who rely on fishing.
2. They impact on human health : the increase of temperatures is conducive to a rise of epidemics and doesn't facilitate sanitary conditions. Therefore it is likely that epidemics such as cholera, malaria, dengue fever will increase in some areas. In the Andes Mountain in Columbia for example, mosquitoes that can carry dengue and yellow fever viruses were previously limited to 3,300 feet (1,006 m) but recently appeared at 7,200 feet (2,195 m). In Mexico as well, dengue fever has spread above its former elevation limit of 3,300 feet (1,006 m) and has appeared at 5,600 feet (1,707 m).
4. They are a threat to peace and security : scarcity of resources leads to competition and could exacerbate tensions already existing between different ethnic groups, countries, etc.
5. They will lead to environmental migrations and will be the source of a new category of refugees or internal displaced populations (IDPs) leaving lands at risk, voluntarily or forced by the governments. Rising sea levels damaging coastal

> regions through flooding and erosion, desertification and shrinking freshwater supplies displaced up to 10 million people in year 2006, and will create up to 50 million environmental refugees by the end of the decade. The Red Cross says environmental disasters already displace more people than war.

Environment-related migration has been most acute in sub-Saharan Africa, but also affects millions of people in Asia and India. Europe and the US face increased pressure from people driven from North Africa and Latin America by deteriorating soil and water conditions. Small island states like Tuvalu are facing the immediate consequences of climate change. New Zealand has already agreed to accept the 11,600 inhabitants of the low-lying Pacific island state if rising sea levels swamp the country. Elsewhere, as many as 100 million people live in areas that are below sea level or liable to storm surge. A total of 213 communities in Alaska are threatened by tides that creep three metres further inland each year.

Climate change has a bigger impact on poor communities, which are the most vulnerable. Developing countries are more vulnerable, due to the economic importance of climatic sensitive sectors (agriculture, fisheries) and their limited human, institutional and financial capacity to anticipate and respond to the direct and indirect effects of climate change. It is likely that this vulnerability is higher for the least developed countries located in the tropical and sub-tropical areas. Often extreme weather events set back the development process for decades.

Major Indicators

Evidence for climatic change is taken from a variety of

sources that can be used to reconstruct past climates. Most of the evidence is indirect—climatic changes are inferred from changes in indicators that reflect climate, such as vegetation, dendrochronology, ice cores, sea level change, and glacial retreat.

Pollen Analysis: Also known as palynology, is based on the notion that the geographical distributions of plant species varies due to particular climate requirements, and that these requirements are the same today as they have been in the past (Uniformitarianism). Each plant species has a distinctively shaped pollen grain, and if these fall into oxygen-free environments (depositional environments), such as peat bogs, they resist decay. Changes in the pollen found in different levels of the bog indicate, by implication, changes in climate.

One limitation of this method is the fact that pollen can be transported considerable distances by wind, wildlife and in some cases running water. Certain depositional sites such as mires may also have been effected by humans through peat cutting for fuel. This has to be taken into consideration when interpretation the pollen record.

Beetles: Remains of beetles are common in freshwater and land sediments. Different species of beetles tend to be found under different climatic conditions. Knowledge of the present climatic range of the different species, and the age of the sediments in which remains are found, allows past climatic conditions to be inferred.

Glacial Geology: Advancing glaciers leave behind moraines and other features that often have datable material in them, recording the time when a glacier advanced and deposited a feature. Similarly, the lack of glacier cover can be identified by the presence of datable ~oil or volcanic tephra horizons. Glaciers are considered one of the most sensitive climate indicators by the IPCC, and their recent observed variations provide a global signal of climate change. See Retreat of glaciers since 1850.

Historical Records: Historical records include cave paintings, depth of grave digging in Greenland, diaries, documentary evidence of events (such as 'frost fairs' on the Thames) and evidence of areas of vine cultivation. Daily weather reports have been kept since 1873, and the Royal Society has encouraged the collection of data since the seventeenth century. Parish records are often a good source of climate data.

Evidence of Climate Change

Climate change has continued throughout the entire history of Earth. The field of paleoclimatology has provided information of climate change in the ancient past, supplementing modern observations of climate.

1. Climate of the deep past
    - Faint young sun paradox
    - Snowball earth
    - Oxygen Catastrophe
2. Climate of the last 500 million years
    - Phanerozoic overview
    - Paleocene–Eocene Thermal Maximum
    - Cretaceous Thermal Maximum
    - Permo–Carboniferous Glaciation
    - Ice ages
3. Climate of recent glaciations
    - Dansgaard–Oeschger event

- o Younger Dryas
- o Ice age temperatures

4. Recent climate
   - o Holocene Climatic Optimum
   - o Medieval Warm Period
   - o Little Ice Age
   - o Year Without a Summer
   - o Temperature record of the past 1000 years
   - o Global warming
   - o Hardiness Zone Migration

There has been a debate about how climate change could affect the world economy. An October 29, 2006 report by former Chief Economist and Senior Vice-President of the World Bank Nicholas Stern states that climate change could affect growth, which could be cut by one-fifth unless drastic action is taken. (*Report's stark warning on climate*) Political advisor Frank Luntz recommended the Bush Administration adopt the term "Climate Change" in preference to global warming, while it worked to discredit the idea of global warming science.

The issue of climate change has entered popular culture since the late 20th century. Science historian Naomi Oreskes has noted that "there's a huge disconnect between what professional scientists have studied and learned in the last 30 years, and what is out there in the popular culture". An academic study contrasts the relatively rapid acceptance of ozone depletion as reflected in popular culture with the much slower acceptance of the scientific consensus on global warming.

Climate Change and Biodiversity

Climate change and biodiversity Some of the most immediate effects of recent climate change are becoming apparent through impacts on biodiversity. The life cycles of many wild plants and animals are closely linked to the passing of the seasons; climatic changes can lead to interdependent pairs of species (e.g. a wild flower and its pollinating insect) losing synchronisation, if, for example, one has a cycle dependent on day length and the other on temperature or precipitation. In principle, at least, this could lead to extinctions or changes in the distribution and abundance of species. One phenomenon is the movement of species northwards in Europe. A recent study by Butterfly Conservation in the UK, has shown that relative common species with a southerly distribution have moved north, whilst scarce upland species have become rarer and lost territory towards the south. This picture has been mirrored across several invertebrate groups. Drier summers could lead to more periods of drought, potentially affecting many species of animal and plant. For example, in the UK during the drought year of 2006 signficant numbers of trees dies or showed dieback on light sandy soils. Wetter, milder winters might impact on temperate mammals or insects by preventing them hibernating or entering torpor during periods when food is scarce. One predicted change is the ascendance of 'weedy' or opportunistic species at the expense of scarcer species with narrower or more specialised ecological requirements. One example could be the expanses of bluebell seen in many woodlands in the UK. These have an early growing and flowering season before competing weeds can develop and the tree canopy closes. Milder winters can allow weeds to overwinter as adult plants or germinate sooner, whilst trees leaf earlier, reducing the length of the window for bluebells to complete their life cycle.

Organisations such as Wildlife Trust, World Wide Fund for Nature, Birdlife International and the Audubon Society are actively monitoring and research the effects of climate change on biodiversity. They also advance policies in areas such as landscape scale conservation to promote adaptation to climate change.

Scientists have confirmed that the earth is warming, and that greenhouse gas emissions from cars, power plants and other manmade sources—rather than natural variations in climate—are the primary cause. Due largely to the combustion of fossil fuels, atmospheric concentrations of carbon dioxide, the principal greenhouse gas, are at a level unequalled for more than 400,000 years. Resultantly, an enhanced greenhouse effect is trapping more of the sun's heat near the earth's surface and gradually pushing the planet's climate system into uncharted territory.

It is significant to note that Carbon dioxide (CO_2) and other greenhouse gases always have been present in the atmosphere, keeping the earth hospitable to life by trapping heat. Yet, since the industrial revolution, emissions of these gases from human activity have accumulated steadily, trapping more heat and exacerbating the natural greenhouse effect. As a result, global average temperatures have risen both on land and in the oceans, with observable impacts already occurring that foretell increasingly severe changes in the future. Polar ice is melting. Glaciers around the globe are in retreat. Storms are increasing in intensity. Ecosystems around the world already are reacting, as plant and animal species struggle to adapt to a shifting climate, and new climate-related threats emerge. Scientists predict that if the increase in greenhouse gas emissions continues unabated, temperatures will rise by as much as 10 degrees Fahrenheit by the end of this century, potentially causing dramatic—and irreversible—changes to the climate. The consequences, both anticipated and unforeseen, will have profound ramifications for humanity and the world as a whole. Water supplies in some critical areas will dwindle as snow and ice disappear. Sea levels will rise, threatening coastal populations. Droughts and floods will become more frequent. And hurricanes and other powerful storms will increase in intensity. Adding to the threat will be the impacts of climate change on agricultural production and the spread of disease. Human health will be jeopardized by all of these changes.

No doubt, limate change is a real problem, but it also has real solutions. Some of its effects are already inevitable and will require some degree of adaptation. But humanity has the power—working collectively and individually and at all levels of society—to take serious action to reduce the threat posed by climate change.

To avoid the worst effects, scientists say we will need to stabilize greenhouse gas concentrations in the atmosphere; that means reducing emissions of these gases by about 50 to 80 percent. It is a major challenge that will require unprecedented cooperation and participation across the globe. Yet, the tools exist to begin addressing this challenge now.

Remarkably, climate change is not just a daunting challenge; it is also an enormous opportunity for innovation. While there is no "silver bullet" technological solution, many tools already exist for addressing climate change, and new options on the horizon could potentially yield reductions in worldwide emissions of greenhouse gases dramatically.

Although greenhouse gas emissions are primarily associated with the burning of fossil fuels (chiefly, coal, oil and natural gas), they come from many sources. As a result, any effort to reduce the human impact on the climate will need to engage all sectors of society. The largest contributors to total U.S. emissions are the electricity generation and transportation sectors; significant emissions also come from other commercial and agricultural activity and from residential and industrial buildings. In each of these areas, technologies and practices already exist that can reduce emissions. Other tools that are still being developed hold tremendous promise. Significant reductions will require a transformation in global energy use through a combination of short-term and long-term commitments.

Real reductions are possible today, but we also need more advanced technology—and we need to begin developing it now. Given the many sources of emissions, a comprehensive response to climate change requires a portfolio of solutions. In the electricity sector, these solutions include improving the efficiency of power plants; generating an increasing share of electricity from climate-friendly renewable sources such as solar, wind and tidal power; developing new technologies to store carbon-dioxide emissions underground; and investing in new nuclear facilities. Increased energy efficiency in buildings and appliances also can provide significant and cost-effective reductions. At the same time, transportation-sector emissions can be reduced through investments in new and existing technologies to improve the fuel efficiency of cars and trucks. Other transportation solutions include using low-carbon fuels and adopting "smart growth" policies that reduce driving. There will certainly be costs associated with adopting these technologies and transforming the way we consume energy.

Yet, addressing climate change also offers enormous economic opportunities, starting with the opportunity to avoid the considerable costs that climate change will pose to societies and businesses.

In addition, the global technology revolution that is needed to protect the climate will create new economic opportunities for businesses and workers, as well as the localities, states and nations that successfully position themselves as centers of innovation and technology development for a low-carbon world. However, innovation will not happen quickly enough or at the necessary scale without government action to push and pull new technologies into mainstream use. Thus needed comprehensive strategy of economy-wide and sector-specific policies. Key policy solutions include investments in science and technology research; efficiency standards for buildings, vehicles, and appliances; and perhaps most importantly, an overall limit on GHG emissions and a market for reductions. One such system, known as cap-and-trade, would set a cap on GHG emissions and allow companies to trade emission allowances so they can achieve their most cost effective reductions.

Remarkably, in the absence of a strong U.S. federal policy, leaders in business and government at all levels have begun taking significant steps to address climate change. Current efforts cannot deliver the level of reduction needed to protect the climate, but they provide a foundation for future action, as well as proof that progress is possible without hampering the current economic success/ achievements.

Leading businesses around the globe are taking action to reduce their impact on the climate and to advocate for sensible policy solutions. A survey of over 30 companies asking why they are taking action on climate change revealed a number of key motivations for action, including increasing profits, influencing government regulation, enhancing corporate reputations, and managing risk.

As a recent development, we have seen a shift in corporate approaches to climate change from focusing exclusively on risk management and protecting the bottom line to the pursuit of new business opportunities. Improvements in energy efficiency, for example, can lead to reduced costs; sales of climatefriendly products and services are growing rapidly; and new markets for carbon reductions are taking off. Many corporate leaders increasingly believe that the growing certainty about climate science means that government action is imminent. Companies want a head start over their competitors in learning how to reduce their emissions. Others in the private sector are responding to growing pressure from investor and consumer groups for disclosure of climate-related risks and integration of climate concerns into companies' core business strategies.

There may also be considerable risk to a company's brand and reputation if customers, partners, investors and/or employees don't view the firm as responsible with regard to climate change. The potential physical impact of climate change on business operations is another concern among corporate leaders of the world.

Recognizing both that government action is inevitable and that policy decisions made on this issue will have substantial implications for future profits, business leaders increasingly are engaging with policymakers to help influence those decisions. Many of these business leaders favor approaches that level the playing field among companies, create more certainty for businesses, and spread responsibility for GHG emission reductions across all sectors of the economy.

The Pew Center on Global Climate Change's Business Environmental Leadership Council includes more than 40 companies at the forefront of corporate action on climate change. Council members' diverse, innovative efforts show the power of business to have a significant impact on reducing GHG emissions while helping the bottom line. These companies employ over 3 million people and have combined a stock market value of over $2.4 trillion. Thirty-two of these companies have set targets that reduce their greenhouse gas emissions.

Thus, obviously, climate change is a global problem requiring a global response. CO_2 emissions have risen 130-fold since 1850 and are projected to increase another 60 percent by 2030. Most emissions come from a relatively small number of countries. An effective strategy to avert dangerous climate change requires commitments and action by all the world's major economies.

It is significant here to mention that the United State of America, with 5 percent of the world's population, is responsible for 25 percent of global GHG emissions, more than any other country. On an intensity basis (emissions per gross domestic product or GDP), U.S. emissions are roughly 50 percent higher than the European Union's or Japan's. On a per capita basis, U.S. emissions are roughly twice as high as those of the EU and Japan (and five times the world average). U.S. emissions are projected to rise 8 percent above 2004 levels by 2010 (and 28 percent by 2025). While by comparison, emissions are projected to hold steady in the EU, and decline 5 percent in Japan, by 2010.

Again, notably missions are rising fastest in developing countries. China's emissions are projected to nearly double, and India's to increase an estimated 80 percent, by 2025. Annual emissions from all developing countries are projected to surpass those of developed countries between 2013 and 2018. Their per capita emissions, however, will remain much lower than those of developed countries. In 1992, countries signed the United Nations Framework Convention on Climate Change with the objective of avoiding dangerous human interference in the climate system (189 countries, including the United States, have ratified the agreement). In the Convention, developed countries agreed to "take the lead" in addressing climate change and to the voluntary "aim" of reducing their emissions to 1990 levels by 2000. Soon recognizing that stronger action was needed, governments launched new negotiations on binding emission targets for developed countries. The resulting agreement, the Kyoto Protocol, requires industrialized countries to reduce emissions on average 5.2 percent below 1990 levels by 2008–2012. All major industrialized countries except the United States and Australia have ratified the protocol.

At the national and regional levels, a range of policies contribute to reducing GHG emissions. The most far-reaching is the European Union's Emissions Trading Scheme, which caps emissions from 12,000 facilities across 25 countries. In major developing countries like China and India, policies driven by economic, energy, or development objectives in many cases contribute to GHG reduction. China, for instance, reduced its energy intensity 68 percent from 1980 to 2000 and has ambitious targets to further improve energy efficiency and expand renewable energy. In 2005, governments launched new processes under the Framework Convention and the Kyoto Protocol to consider next steps in the international effort.

The report of the Climate Dialogue at Pocantico, a group of senior policymakers and stakeholders from 15 countries convened by the Pew Center on Global Climate Change, calls for a flexible international framework allowing different countries to take on different types of commitments (including economy-wide emission targets, sectoral agreements, and policy-based approaches).

The future of the international effort hinges in large measure on the United States—other major emitters are unlikely to commit to stronger action without the participation of the world's largest economy and emitter. As it strengthens its domestic response to climate change, the United States must also provide the leadership needed for an effective long-term global effort.

In February 2002, US President Bush announced a voluntary target to achieve an 18-percent reduction in U.S. greenhouse gas intensity (the ratio of emissions to gross domestic product) by 2012. Under this target, emissions actually will continue to rise as the economy grows. In 2004, U.S. emissions were 18 percent higher than they were in 1990, and 2.6 percent higher than at the start of 2002. A number of senators and representatives—both Democrats and Republicans—have offered proposals to limit emissions, but a mandatory climate bill has yet to pass in either branch of Congress. Nonetheless, momentum for action is growing, as indicated by the increasing number of bills, votes and hearings held on climate-related issues in Congress in recent years.

The lack of action in Washington on the climate issue has prompted many states to seek their own solutions both individually and cooperatively. At this point, nearly every US state is engaged in working in some way on climate solutions. By taking action to address climate change U.S. states are fulfilling their role in American democracy as "policy laboratories," developing initiatives that serve as models for federal action.

Till now twenty-eight states have adopted climate action plans detailing steps they will pursue in addressing climate change, and 12 states actually have set targets, ranging from modest to aggressive, to reduce their GHG emissions in the decades ahead. Beyond these broad-based plans and targets, many states have adopted sector- specific policies that reduce emissions from electricity generation —for example, by promoting the development of clean and renewable energy resources and by requiring that utilities generate a specified share of power from renewable sources. States also are directing public funds to energy efficiency and renewable energy projects and adopting new standards for power plant emissions and energy efficiency. In the transportation sector, states are adopting policies and standards to promote efficient, low-emission vehicles and climate-friendly fuels. They are also working on smart growth, zoning reform, and transit-oriented development.

Agricultural policies also are being redesigned to promote biomass as another solution to climate change.

Among the main motivating factors for state action has been concern about the potential impact of climate change on state economies from consequences such as sea level rise or extreme weather. However, many state leaders also see enormous and largely untapped economic opportunities that will come with developing new markets for climate-friendly technologies.

In contrast to the global warming debate at the federal level, climate-related policies typically enjoy bipartisan support among the states. This activity on the part of states is significant because some U.S. states are major emitters of greenhouse gases, producing levels at par with those of many developed countries. In addition, state actions are showing it is possible to reduce emissions and spur technological innovation without endangering economic competitiveness. And, through interstate partnerships, states are demonstrating the power of collective action to reduce costs and to achieve increased efficiency, while cutting emissions across a larger geographic area.

Besides spotlighting what works, however, states also are demonstrating that their efforts alone are not enough. States have limited resources and strict budget requirements that make far-reaching climate policies difficult to implement, and they also lack certain powers that would be crucial to a comprehensive climate change policy.

Moreover, the patchwork quilt that can result when states take individual approaches to the climate issue can be inefficient and pose challenges for business. State action is important, but strong and coherent federal policies are needed to ensure consistency and to mobilize climate solutions throughout the economy and the nation.

Remarkably, State leaders are hardly alone in their movement to address climate change. Across the country and all over the world, city and county governments are implementing their own policies aimed at reducing greenhouse gas emissions. Cities have a strong history of climate action, and continue to mount responses to climate change that are resulting in emissions cuts. Cities are working together to achieve their goals through a number of programs and mechanisms, including the International Council for Local Environmental Initiatives and the U.S.

Mayors Climate Protection Agreement, both of which have experienced dramatic growth in participation. Policies adopted by cities and towns within USA span everything from energy supply to transportation to tree planting. Local leaders are taking action because they recognize that their communities have a lot to lose should emissions remain unchecked and climate change accelerate. Many of the potential effects of climate change—such as extreme weather, higher sea levels and reduced water supplies — will be felt most sharply by urban populations. In addition to reducing risks, cities and towns also can realize indirect benefits by tackling climate change, such as energy savings and improved air quality. Localities, like the states, are offering lessons in what works to protect the climate.

Climate change is happening, and the time to act is now. While the early actions of local and state governments, nations, and business leaders are significant, climate change remains a global problem requiring a global solution. Ultimately, a fair and effective international approach must engage all of the world's major economies and allow enough flexibility for all countries to contribute.

Causes of Global Climate Change

During the twentieth century, the earth's surface warmed by about 1.4 °F. There are a variety of potential causes for global climate change, including both natural and human-induced mechanisms. Science has made great strides recently in determining which potential causes are actually responsible for the climate change that occurred during the twentieth century, providing strong evidence that greenhouse gases released to the atmosphere by human activities are the main cause of contemporary global warming.

Recent decades have seen record-high average global surface temperatures. Thermometer readings sufficient to provide reliable global averages are available back to 1850 (Brohan et al. 2006). In the past quarter-century, according to satellite measurements, the lower atmosphere warmed by 0.22- 0.34 °F per decade, equivalent to 2-3 °F per century (Christy and Spencer 2005; Mears and Wentz 2005). The past 20 years include the 18 warmest years on record (Hadley Centre, 2005).

This well-documented warming trend could result from several factors that influence the earth's climate, some of which are natural, such as changes in solar radiation and volcanic activity. Others, particularly the release of certain gases to the atmosphere and land-cover changes, are man-made. This brief describes recent scientific progress in identifying the causes.

The greenhouse effect is a natural phenomenon whereby certain gases in the earth's atmosphere, known as greenhouse gases, absorb heat that would otherwise escape to space. This heat originates from visible sunlight that warms the earth's surface. Subsequently, heat radiates from the surface to the atmosphere, where some of it is absorbed by greenhouse gases and radiated back to the surface. Recent progress in climate modelling has generated a consensus among climate scientists that greenhouse gases emitted by human activities are likely to have caused most of the observed global temperature rise over the past 50 years (Mitchell et al. 2001). The increase in the strength of the greenhouse effect as a result of man-made greenhouse gases is known as the enhanced greenhouse effect.

Global climate varies over time in response to climate forcings—physical factors external to the climate system that force a net increase (positive forcing) or net decrease (negative forcing) of heat in the climate system as a whole (Hansen, Sato et al. 2005). This type of change is distinct from internal climate variability, in which heat is transported by winds or ocean currents between different components of the climate system with no net change in the total heat within the system. The El Niño–Southern Oscillation is a well-known example of internal climate variability. Because the observed climate change over the twentieth century results from a net increase of heat in the entire climate system, it can only be explained by external forcing (Hansen, Nazarenko et al. 2005).

It must be remembered here that until recent centuries, climate forcings were exclusively natural, such as changes in the amount of sunlight reaching the earth's surface and changes in emissions of dust from volcanoes. During modern times, human activities have introduced a mix of additional forcings, such as increases in atmospheric greenhouse gases that cause warming (positive forcing), and sulfate aerosols, miniscule particles that reflect sunlight and cause cooling (negative forcing). The histories and magnitudes of various forcings are estimated from direct observations, such as satellite measurements of solar radiation in recent decades, or from proxies, such as sunspots for solar radiation in earlier decades (Foukal et al. 2004).

Fingerprint matching between climate forcings and observed climate change is performed using physical climate models that calculate how each forcing should have affected climate over time, based on its history and how scientists understand the physical mechanisms of each forcing. These models are able to reproduce most of the major features of the global climate system, including the pattern of global warming over the past century.

The models serve as controlled experiments that test alternative hypotheses about the causes of climate change. Entering records of one or more forcings into a model, scientists assess whether the climate scenario generated by the model is similar to the observed climate record; no observed climate data are entered into the model. If the simulated climate matches observed climate, then the forcing(s) represented in the model can explain the observed climate record. If not, the forcing(s) cannot explain the observed climate change. Of course, it is possible that more than one forcing is involved, so scientists test all possible combinations of forcings to see if their combined influence can explain observed climate change.

It is significant here to note that the twentieth-century warming trend at the earth's surface progressed in a distinct pattern, with a large warming during 1910-1940, moderate cooling during 1940-1975, and a large warming from 1975 to the present. Scientists at the National Center for Atmospheric Research (NCAR) looked for fingerprints of various natural (solar radiation, volcanic particles) and man-made (greenhouse gases, sulfate aerosols) forcings in this record of observed climate change (Meehl et al. 2004).

The study employed a physical climate model that allowed individual or combinations of forcings to drive the simulated climate. The change in surface temperature calculated by the model for each forcing or combination of forcings was then compared with the observed record of surface temperature change over the twentieth century. The best fit of the model results to the observed climate was produced when all of the forcings were included, implicating all of the forcings in producing the overall pattern of change.

However, different forcings dominated at different times

during the century. For instance, the temperature rise in the early part of the 20th century was dominated by natural forcings, whereas the warming after 1975 was dominated by man-made greenhouse gases. The cooling during the mid-century was consistent with a combination of natural volcanic and man-made aerosols. The results of this study implicate the enhanced greenhouse effect as the dominant cause of global warming over the past three decades. If not for the temporary cooling between 1940 and 1975 from volcanic and man-made aerosol emissions, the earth might be even warmer than it is today.

Oceans exhibit natural temperature cycles, with some oceans cooling at the same time that others warm. This natural internal variability of climate results from heat radiation from one place to another, but it adds no new heat to the ocean as a whole. A major challenge for assigning a cause to temperature changes is distinguishing internal variability from external forcing, which adds new heat to the system. Recently, scientists from the U.S. National Oceanographic and Atmospheric Administration demonstrated that the ocean as a whole has been warming for the past five decades (Levitus et al. 2005).

The first principles of physics dictate that simultaneous warming of all the world's oceans could only occur through external forcing, as there is no other source of this much energy within the climate system (Hansen, Nazarenko et al. 2005). Using a fingerprinting-modeling approach similar to the one described above, scientists at Scripps Institution of Oceanography, Lawrence Livermore National Lab, NCAR, and the United Kingdom's Hadley Center, published a study showing that the oceans situated along the equator have warmed over the past five decades as a direct result of the enhanced greenhouse effect (Barnett et al. 2005).

Notably, Observations show that the oceans have been warming from the surface downward, which indicates heat transfer from the atmosphere. The vertical pattern of heat penetration with depth varies from ocean to ocean as a result of internal variability (i.e. currents transporting heat from one ocean to another). This complex pattern of vertical profiles provides a "fingerprint" of climate forcing. Modeling of internal variability alone or internal variability combined with solar and volcanic forcings did not produce temperature profiles that matched this fingerprint. However, the combined influence of human-induced forcings, natural forcings, and internal variability reproduced the pattern of heat penetration for each ocean. Man-made greenhouse gases strongly dominated the overall forcing.

Another fingerprint of the enhanced greenhouse effect has been identified in the observed increase in the height of the tropopause, a region of the earth's atmosphere that represents the transition between the lower atmosphere (troposphere) and the upper atmosphere (stratosphere). Factors that either warm the troposphere or cool the stratosphere increase the tropopause elevation, and climate models have long predicted that the elevation of the tropopause above the earth's surface should increase as a result of the enhanced greenhouse effect (Santer et al. 2003).

Although this phenomenon may affect climate behavior, it is discussed here strictly as a tool for identifying causes of observed climate change. Scientists from several American, British, and German research institutions employed a fingerprinting-modeling approach to determine which climate forcings could explain observed changes in the height of the tropopause (Santer et al. 2003, 2004). Between 1979 and 2001, satellites monitoring the atmosphere recorded a

620-foot rise of the tropopause. In the model simulations forced by both natural and human-induced forcings, the tropopause elevation increased similarly. Manmade greenhouse gases, which warmed the troposphere, and stratospheric ozone depletion (by man-made chemicals), which cooled the stratosphere, dominated the forcing. Manmade greenhouse gases caused about 40 percent of the rise, whereas ozone depletion caused about 60 percent. Overall, the effect of solar forcing, which contributed slightly (less than 10%) to the rise of the tropopause, was canceled by a small negative forcing (decrease in tropopause height) from volcanoes. Thus, human-induced forcings from greenhouse gases and ozone-depleting chemicals provide the best explanation for the observed increase in the elevation of the tropopause.

Natural internal variability is an inherent feature of the climate system, but it cannot account for the net gain of energy that has been detected within the climate system as a whole. Based on physical principles, the modern increase in the heat content of the global ocean demonstrates that positive external forcing of the climate is underway.

Changes in natural external forcings cannot explain the observed global warming of recent decades. Records of observed climate change at the earth's surface, in the global ocean, and in the atmosphere, bear the fingerprint of the enhanced greenhouse effect, which is caused by human activities associated with fossil fuel burning and land use.

CLIMATE CHANGE AND AGRICULTURE

Global warming is projected to have significant impacts on conditions affecting agriculture, including temperature, precipitation and glacial run-off. These conditions determine the carrying capacity of the biosphere to produce enough food for the human population and domesticated animals. Any short-term fluctuations of the climate can have dramatic effects on the agricultural productivity. Thus, the climate has a direct influence on food supply. Rising carbon dioxide levels would also have effects, both detrimental and beneficial, on crop yields. A positive effect of rising carbon dioxide concentrations would be increased agricultural yields, because of the role of carbon dioxide in photosynthesis. The overall effect on agriculture will depend on the balance of the beneficial and detrimental effects.

In spite of technological advances, such as improved varieties or HYV, genetically modified organisms, and irrigation systems, weather is still a key factor in agricultural productivity, as well as soil properties and natural communities. The effect of climate on agriculture is related to variabilities in local climates rather than in global climate patterns. Consequently, agronomists consider any assessment has to be individually consider each local area.

On the other hand, agricultural trade has grown in recent years, and now provides significant amounts of food, on a national level to major importing countries, as well as comfortable income to exporting ones. The international aspect of trade and security in terms of food implies the need to also consider the effects of climate change on a global scale.

IPCC

The 2001 IPCC Third Assessment Report concluded that the poorest countries would be hardest hit, with reductions in crop yields in most tropical and sub-tropical regions due to decreased water availability, and new or changed insect pest incidence. In Africa and Latin America many rainfed crops are near their maximum temperature tolerance, so that yields are likely to fall sharply for even small climate changes; falls in agricultural productivity of up to 30% over the 21st century are projected. Marine life and the fishing industry will also be severely affected in some places.

Shortage in Grain Production

Between 1996 and 2003, grain production has stabilized slightly over 1800 millions of tons. In 2000, 2001, 2002 and 2003, grain stocks have been dropping, resulting in a global grain harvest that was short of consumption by 93 millions of tons in 2003.

The earth's average temperature has been rising since the late 1970s, with the three warmest years on record coming in the last five years. In 2002, India and the United States suffered sharp harvest reductions because of record temperatures and drought. In 2003 Europe suffered very low rainfall throughout spring and summer, and a record level of heat damaged most crops from the United Kingdom and France in the Western Europe through Ukraine in the East. Bread prices have been rising in several countries in the region.

Increases in Agricultural Production

Between 2003 and 2004 worldwide wheat production increased 13%, oilseed production increased 16%, rice production increased 3%, and cotton production increased 23%. although this may not be directly related to global warming.

Some major limitations to climate changes consequences estimates are related to the models that are being used. The *climate models* do not have a true ability to give accurate projections because of inadequate understanding of natural processes and the limitation of computing power, and the sheer amount of variables which bring in the need to invoke chaos theory. As a consequence, the assessment of possible effects of climate changes are based on estimations. Most models are also not able yet to provide reliable projections of changes in climate variability on a local scale, or in frequency of exceptional events such as storms and drought. For example, there tends to be a lack of consensus among experts in prediction of regional soil moisture changes.

Crop Development Models

In order to further study effects of global warming on agriculture, other types of models, such as *crop development models*, *yield prediction*, quantities of *water or fertilizer consumed*, can be used. Such models condense the knowledge accumulated of the climate, soil, and effects observed of the results of various agricultural practices. They thus could make it possible to test strategies of adaptation to modifications of the environment.

Because these models are necessarily simplifying natural conditions, it is not clear whether the results they give will have an *in-field* reality. However, some results are partly validated with an increasing number of experimental results.

Other Types of Biological Models

Other models, such as *insect and disease development* models based on climate projections are also used (for example simulation of aphid reproduction or septoria (cereal fungal disease) development).

Scenarios are used in order to estimate climate changes effects on crop development and yield. Each scenario is defined as a set of meteorological variables, based on generally accepted projections. For example, many models are running simulations based on doubled carbon dioxide projections, temperatures raise ranging from 1°C up to 5°C, and with rainfall levels an increase or decrease of 20%. Other parameters may include humidity, wind, and solar activity. Scenarios of crop models are testing farm-level adaptation, such as sowing date shift, climate adapted species (vernalisation need, heat and cold resistance), irrigation and fertilizer adaptation, resistance to disease. Most developed models are about wheat, maize, rice and soybean.

Consequences of Potential Global Climate Changes on Agricultural Production

The first direct effect of climate change is the composition of the earth atmosphere, such the amount of carbon dioxide and ozone. Gases such as methane, nitrogen dioxide and chlorofluorocarbons, however, are commonly believed not to have any effect on physiological processes. Some indirect effects are climate parameters resulting from climate change, such as temperature, insolation, rainfall, and humidity. Other indirect effects include the side effects due to the climatic changes, such as the increase of the sea level (SLR), changes in ocean currents, or tornadoes. All these influences may combine nègatively or positively—the assessment of these effects depends on whether one considers annuals crops (cereals and legumes) or herbaceous perennial cultures (fodder, meadows) or other cultures such as vine or fruit trees. The effects are also different depending on the latitude. In temperate countries, effects are found less negative or even rather beneficial, while in tropical and desertic countries they tend to be adverse. Effects also depend on altitude, for example, places at higher altitudes tend to benefit from a warmer temperature.

Climate change induced by increasing greenhouse gases is likely to affect crops differently from region to region. *For example, average crop yield is expected to drop down to 50% in Pakistan according to the UKMO scenario whereas corn production in Europe is expected to grow up to 25% in optimum hydrologic conditions.*

More favourable effects on yield tend to depend to a large extent on realization of the potentially beneficial effects of carbon dioxide on crop growth and increase of efficiency in water use. Decrease in potential yields is likely to be caused by shortening of the growing period, decrease in water availability and poor vernalization.

Duration of crop growth cycles are above all, related to temperature. An increase in temperature will speed up development. In the case of an annual crop, the duration between sowing and harvesting will shorten (for example, the duration in order to harvest corn could shorten between one and four weeks). The shortening of such a cycle would have an adverse effect on productivity because senescence would occur sooner.

Carbon dioxide is essential to plant growth. Rising CO_2 concentration in the atmosphere can have both positive and negative consequences.

Increased CO_2 is expected to have positive physiological effects by increasing the rate of photosynthesis. Currently, the amount of carbon dioxide in the atmosphere is 380 parts per million. In comparison, the amount of oxygen is 210,000 ppm. This means that often plants may be starved of carbon dioxide, being outnumbered by the photosynthetic pollutant oxygen. The effects of an increase in carbon dioxide would be higher on C_3 crops (such as wheat) than on C4 crops (such as maize), because the former is more susceptible to carbon dioxide shortage. Under optimum conditions of temperature and humidity, the yield increase could reach 36%, if the levels of carbon dioxide are doubled.

Plants take in CO_2 through multiple tiny pores in their leaves called stomata. These pores can open or close in response to the plant's requirements. A higher ambient level of carbon dioxide allows plants to respire CO_2 more readily with less stomatal opening. Plants lose water due to transpiration through open stomata, so they are adapted to conserve water by opening the stomata only as necessary.

The stomata also serve to allow the plant to eliminate oxygen which is the by-product of photosynthesis. Higher carbon dioxide levels within cells can allow the stomata to be closed longer without suffering photorespiration, an effect of too much oxygen in ratio to carbon dioxide in the plant cell's chloroplasts.

However, other studies also show a change in harvest quality. The growth improvement in C3 plants could favor vegetative biomass on grain biomass; thus leading to a decrease in grain production yield.

Carbon dioxide is believed by many scientists to be potentially responsible of increase in agricultural production: a 10-15 % increase for wheat and soybean, 8% for corn and rice for a +2°C scenario on average. However, these results disregard great differences among countries, not least between varying climates.

According to the IPCC's TAR, "The importance of climate change impacts on grain and forage quality emerges from new research. For rice, the amylose content of the grain—a major determinant of cooking quality—is increased under elevated CO_2" (Conroy et al., 1994). Cooked rice grain from plants grown in high-CO_2 environments would be firmer than that from today's plants. However, concentrations of iron and zinc, which are important for human nutrition, would be lower (Seneweera and Conroy, 1997). Moreover, the protein content of the grain decreases under combined increases of temperature and CO_2.

Studies have shown that higher CO_2 levels lead to reduced plant uptake of nitrogen (and a smaller number showing the same for trace elements such as zinc) resulting in crops with lower nutritional value. This would primarily impact on populations in poorer countries less able to compensate by eating more food, more varied diets, or possibly taking supplements.

Reduced nitrogen content in grazing plants has also been shown to reduce animal productivity in sheep, which depend on microbes in their gut to digest plants, which in turn depend on nitrogen intake.

Global Warming and Water Distribution

Global warming would be able to modify the global distribution of water, possibly leading to several effects, both detrimental and less beneficial.

If global warming happens, many believe the general ability to predict weather patterns will decrease, due to more extreme weather. This would make it more difficult to plan agricultural actions. If extreme climatic conditions become more frequent, there would be more intense rainfall, droughts and heat spells in different parts of the globe or the year.

Climate change is likely to increase the amount of arable land near the poles by reduction of the amount of frozen lands. Sea levels are expected to get up to one meter higher by 2100, though this projection is disputed. Rise in sea level would result in agricultural land loss, in particular in areas such as South East Asia. Erosion, submergence of shorelines, salinity of the water table due to the increased sea levels, could mainly affect agriculture through inundation of low-lying lands.

Soil Erosion and Fertility

With global warming, soil degradation is more likely to occur, and soil fertility would probably be affected by global warming. However, due to the fact that the ratio of carbon to nitrogen is a constant, a doubling of carbon is likely to imply a higher storage of nitrogen in soils as nitrates, thus providing higher fertilizing elements for plants, providing better yields. The average needs for nitrogen could decrease, and give the opportunity of changing often costly fertilisation strategies.

Due to the extremes of climate that would result, the increase in precipitations would probably result in greater risks of erosion, according to the intensity of the rain. The possible evolution of the organic matter in the soil is a highly contested issue: while the increase in the temperature would induce a greater rate in the production of minerals, lessening the soil organic matter content, the atmospheric CO_2 concentration would tend to increase it.

Potential Effects of Global Climate Change on Pests, Diseases and Weeds

A very important point to consider is that weeds would undergo the same acceleration of cycle as cultivated crops, and would also benefit from carbonaceous fertilization. Since most weeds are C_3 plants, they are likely to compete even more than now against C_4 crops such as corn. However, on the other hand, some results make it possible to think that weedkillers could gain in effectiveness with the temperature increase.

Global warming would cause an increase in rainfall in some areas, which would lead to an increase of atmospheric humidity and the duration of the wet seasons. Combined with higher temperatures, these could favor the development of fungal diseases. Similarly, because of higher temperatures and humidity, there could be an increased pressure from insects and disease vectors.

Some scientists think agriculture could be affected by any decrease in stratospheric ozone, which could increase biologically dangerous ultraviolet radiation B. Excess ultraviolet radiation B can directly affect plant physiology and cause massive amounts of mutations, and indirectly through changed pollinator behavior, though such changes are difficult to quantify. However, it has not yet been ascertained whether an increase in greenhouse gases would decrease stratospheric ozone levels.

In addition, a possible effect of rising temperatures is significantly higher levels of ground-level ozone, which would substantially lower yields.

In the long run, the climatic change could affect agriculture in several ways :

- *productivity*, in terms of quantity and quality of crops
- *agricultural practices*, through changes of water use (irrigation) and agricultural inputs such as herbicides, insecticides and fertilizers
- *environmental effects*, in particular in relation of frequency and intensity of soil drainage (leading

to nitrogen leaching), soil erosion, reduction of crop diversity

- *rural space*, through the loss of previously cultivated lands, land speculation, land renunciation, and hydraulic amenities.

They are large uncertainties to uncover, particularly because there is lack of information on many specific local regions, and include the uncertainties on magnitude of climate change, the effects of technological changes on productivity, global food demands, and the numerous possibilities of adaptation.

Most agronomists believe that agricultural production will be mostly affected by the severity and pace of climate change, not so much by gradual trends in climate. If change is gradual, there will be enough time for biota adjustement. Rapid climate change, however, could harm agriculture in many countries, especially those that are already suffering from rather poor soil and climate conditions, because there is less time for optimum natural selection and adaption. The adoption of efficient new techniques tends to be far from obvious. Some believe developed nations are too well-adapted to the present climate. developing nations also would often have extensive social or technical constraints that prevent them from achieving sustainable production.

Since major effects from climate change now appear inevitable, it is vital for countries and communities to take practical steps to protect themselves from the likely disruption and damage that will result. This is known in international jargon as adaptation. In some cases, it's hard to know what to do — how do you stop encroaching oceans? — and so strategies and technologies have to be developed. But the greater difficulties are psychological and political.

Prevention calls for overriding human nature and politicians' instincts. It is difficult to take expensive and painful measures when — in any given place — people don't know exactly what's coming, exactly when, and how bad it will be.

For example, flood walls should be built and in numerous cases it is probably advisable to move human settlements out of flood plains and other low-lying areas, but how high should the walls be and where should people be moved? Even in the case of much more specific threats, it takes brave public officials and stoic populations to act in advance of problems, even though that almost always works better than reacting desperately in times of crises. Few practical precautions against climate change have been taken to date, although the need is urgent. Efforts are at least going forward in the form of assessments and studies. . . measures that are politically palatable and possible at the moment.

One approach that has proven valuable is to focus on vulnerability rather than on the still-uncertain extent of the danger posed by climate change. It is vital, in other words, to know who is going to need help, whatever climate difficulties arise. Further rational decisions can be made and plans developed based on this important information. A number of countries also are carrying out research. Over the last decade, for example, predictions of the impacts of climate change have grown much more detailed and accurate. As that process advances, and as efficient methods for adapting are devised, it will become more politically and economically feasible to take specific precautions.

As a positive development, in 1997, under the United Nations Framework Convention on Climate Change, new funds were set up to support adaptation. A Special Climate Change Fund was established in 2001 to finance projects relating to adaptation. A Least Developed Countries Fund is intended to enable the world's poorer nations to carry out "national adaptation programmes of action." And an Adaptation Fund was established under the Kyoto Protocol to support the implementation of concrete adaptation projects and programmes. Flexibility is a valuable tool. It is a sensible way of planning ahead in the agricultural sector, for example, to grow a variety of crops, some of which may be viable in times of climate flux, rather than investing in a single crop that may be felled by a drought or a heat wave.

It is useful to determine which crops will be profitable and will grow well in warmer temperatures or a drier climate, if that is what is predicted for a given region. The European Union has encouraged farmers to practice flexibility in land use and crop production. It also has suggested modifying subsidy and incentive programmes to influence farm operation and management in ways that will better cope with climate change. Rational decisions about the safe location of new facilities and infrastructure is a valuable, cost-effective step governments and industries can take now and in coming years. It is better to put future electrical-generating plants, train lines, major roads, and hospitals, not to mention houses, in spots where more intense storms and floods won't damage or destroy them.

Rmarkably, one benefit of the generally difficult process of adaptation is that some measures will have positive effects regardless of what happens with climate change. Restoring tree cover, wetlands, and grasslands to prevent erosion and to lessen damage from storms and floods will help people even when normal storms occur — and also will provide sanctuary for wildlife and have aesthetic and recreational benefits. Similarly, setting up evacuation plans and medical response systems for severe storms and floods can save lives in case of eventuality.

Expanding Forests

Forests are a major ally in combating global warming. Trees and other green plants, using only sunlight for energy, take carbon dioxide out of the atmosphere, releasing oxygen and storing carbon in a safe and useful way. Forests, which provide all kinds of undervalued benefits for mankind, can be major allies in the battle against climate change and global warming. . . if only humans start planting them and stop cutting them down (i.e. deforestation).

"Sinks" is the term used by climatologists for vast swaths of trees and other green vegetation which "drain away" the most dominant greenhouse gas.

Deforestation, which is occurring all over the world, has a doubly damaging effect: it reduces the number of trees that can absorb the carbon dioxide produced by human activities, and it releases into the atmosphere the carbon contained in the trees that are cut down.

No doubt, he world at large currently doesn't "pay" much for the positive effects of forests. The value of trees as lumber and as firewood, and the value of the land they occupy for housing or farming, tend to be short-term and specific.

In fact, these benefits may be a matter of survival in some regions. The value of forests for preventing global warming and preserving the earth's biodiversity, by contrast, are long-term and their rewards apply to everyone generally.

A way has to be found to make the expansion and nurturing of forests appealing and cost-effective to the local populations that usually decide their fate. In terms of efforts to reduce global warming, a forest in one place is as good as a forest in another. That can give rise to certain practical arrangements and efficiencies.

Under the Kyoto Protocol, once it takes effect, industrialized countries which lack space or cost-effective options for expanding forests on their own territories may partially compensate for their greenhouse-gas emissions by paying for the establishment and maintenance of forests in other countries regions.

Carbon stored in agricultural soils often can be preserved or enhanced by switching to "no-tillage" or "low-tillage" techniques, which slow the rate at which organic soil matter decomposes. In paddy fields, emissions of methane, a powerful greenhouse gas, can be suppressed to some extent through tillage practices, water management, and crop rotation. Whereas using nitrogen fertilizers more efficiently can reduce emissions of nitrous oxide.

People don't necessarily make decisions based on efficiency or the health of the environment. They are apt to do what they've done in the past, what is expected, what their friends and neighbors do, what is fashionable. Driving a car with a large engine when a small engine — which burns less fuel — will do the job is a matter of choice. In choosing cars and electrical appliances and methods for heating and cooling their homes, human beings don't necessarily think about climate change. And when thousands and even millions of people make choices that add unnecessarily to the global warming problem, the effects can be considerable. Traditions and habits also may limit the lifestyle choices that are available.

Businesses and governments tend not to supply products, services, and policies people don't want. Mass transit is much less wasteful of fossil fuels than automobile use, but if the public hasn't demanded mass transit and the necessary train lines and subway systems and bus routes haven't been built, then they aren't quickly available when and if people change their minds. Momentum has to be built up for such changes . . . and yet, paradoxically, if the options aren't available, it's hard to establish momentum for them.

"Hybrid" cars use roughly half the gasoline of standard automobiles, but because few customers have asked for such vehicles, the price per car is high and few auto makers produce them. Lack of demand also slows innovation and technological progress, while heightened interest accelerates it.

If customers and citizens, especially in industrialized countries, become strongly interested in combating global warming and are willing to buy products that cut emissions, new processes and technologies could well be invented that would greatly reduce or even solve the problem.

As is widely accepted laws and regulations can have a major impact on greenhouse-gas emissions because they affect business behavior and public habits. While some governments encourage the use of mass transit; some — through tax arrangements, road-building programmes, and even subsidies — encourage the burning of fossil fuels. One way of changing behavior is to make it illegal. Another is to make it more expensive through taxes or penalties.

Whereas, some governments, spurred by membership in the United Nations Framework Convention on Climate Change, have already attempted to cut greenhouse gas emissions with a mixture of carrots and sticks — with inducements, subsidies, voluntary programmes, regulations, and fines. While several have attacked the problem directly by imposing "taxes" on carbon use. Others have established "carbon markets" where units of energy use may be bought and sold. These arrangements anticipate provisions that will apply to governments that have ratified the Kyoto Protocol, once the Protocolis implemented.

Minimum standards for energy efficiency in new buildings were updated recently in a series of countries, including Austria, France, Japan, New Zealand, and the United Kingdom. Such measures can include requirements for walls and roofs that limit heat loss. And they can require a minimum level of thermal efficiency for furnaces and water heaters.

Further, standards for energy efficiency in electrical appliances have been established by some governments. A programme begun by Japan in 1998 is expected to cut the energy requirements of home video recorders by 59 per cent, refrigerators by 30 per cent, and computers by 83 per cent. The sting of more stringent regulations can be softened by inducements, as in the Netherlands, where tougher European Community appliance standards are now in effect. In the Netherlands, householders are entitled to a partial rebate on the purchase prices of the most efficient appliances. Economic and fiscal instruments have been used to spur shifts in freight transport from roads to rail systems and ships, which use less fuel per ton of goods transported. Examples are the introduction of road tolls in Austria and mileage-based tolls for lorries in Austria, Germany, and Slovenia; increased investments in rail systems in Austria and Belgium; and promotion of ship and rail use for freight in Belgium, Switzerland, and Japan. Among voluntary arrangements is an agreement reached between the

European Union and European and Asian automobile manufacturer organizations setting reduced targets for carbon-dioxide emissions from passenger cars and light commercial vehicles.

"Green tariffs" have been used by Belgium, Germany, Hungary and Switzerland to spur the use of renewable energy. The "tariffs" guarantee electrical-generating companies a higher price for a unit of renewable energy than the prevailing market price. Also steps have been taken by a number of governments to require waste firms and landfills to prevent the escape of greenhouse gases such as methane. Landfill taxes per ton of waste in Switzerland and Norway are higher for facilties that are not sealed.

Between 1990 and 2000, the total greenhouse-gas emissions of industrialized countries actually declined slightly (by 5.6 per cent) -- but that reflected unusual circumstances. Because of the steep and painful drop in economic output of the countries of Eastern Europe and the erstwhile Soviet Union, which were shifting from centrally planned to market economies, emissions in those nations fell by 37 per cent.

That more than compensated for an 8.2 per cent increase in emissions among developed countries elsewhere. But now the Eastern European economies in transition are growing again, and their emissions are rising. The *real work*, what is *really* needed — worldwide economic progress combined with reduced emissions — has yet to be accomplished. It will not be easy. The good news is that technology already exists that could stabilize and even reduce greenhouse gas levels within a few decades. The problem is that getting such technology in place — installing and paying for more efficient procedures for burning fossil fuels and for using renewable energy sources such as solar power and wind power — is politically and economically difficult. Economic systems and governments currently aren't arranged for accomplishing greenhouse-gas reductions. The costs of cutting emissions tend to be immediate and specific. The benefits, such as fewer severe storms, floods, and droughts, not to mention and a cleaner and safer environment, will occur gradually in the future and will benefit people everywhere, whether they pay for the relevant technology or not. No doubt, it is hard to put a price on these positive outcomes. The system has to be adjusted so that the burdens as well as the benefits are fairly distributed.

Surprisingly, there also are competitiveness problems: if laws and regulations around the world aren't equally demanding, businesses in countries that don't require greenhouse-gas reductions will be able to operate more cheaply and sell their products at lower prices (at least in the short term) than businesses in countries that require more climate-friendly behavior. Multinational corporations may (MNCs) shift their factories to places where regulations are less restrictive.

Some technologies and policies, called "no-regret options," can reduce emissions and pay for themselves through greater efficiency and lower operating costs. . . but even those require investments up front which some businesses may not be able or willing to pay, especially if they don't receive credits or loans from governments. While useful technology may be bought and shared, in the end "no regrets" methods won't be enough to stabilize or reduce worldwide greenhouse-gas levels — governments, businesses, and people are going to have to make difficult choices and take painful steps.

For example Denmark stabilized emissions between 1990 and 2000 by switching to more efficient methods of electricity generation and by shifting from coal use in industry to renewable energy and natural gas.

Similarly, Germany achieved a considerable reduction in carbon dioxide emissions through the economic restructuring resulting from the reunification of the country, through the reduced use of lignite, and through greater use of wind energy. There also were substantial cuts in methane emissions from coal production, waste management, and agriculture, and in nitrous oxide emissions from agriculture. The United Kingdom established a "national climate change" policy that set targets for using energy more efficiently, and was able to lower emissions, among other things by using less coal and more natural gas.

The developing countries are also making attempts within the reach of their resources to abate green house gases. No doubt, their share in GHGs emission is far lesser than the developing countries, they are the worst sufferer (even though potential) of effects of global warning and consequented climate change.

4

CLIMATE CHANGE IN ASIA

> Whenever you are to do a thing, though it can never be known but to yourself, ask yourself how you would act were all the world looking at you, and act accordingly.
>
> —*Thomas Jefferson*

Admittedly, Asia is the world's largest and most populous continent which covers about 30% of the land area of the world and contains more than 60% of the current human population on the earth. The Asian continent is bounded on the north by the Arctic Ocean, on the east by the Pacific Ocean, and on the south by the Indian Ocean; the western boundary with Europe, runs roughly north-south along the eastern Ural Mountains, the Zhem River, the Caspian Sea, the Kuma-Manych Depression, the Black Sea, the Aegean Sea, the Mediterranean Sea, the Suez Canal, and the Red Sea. The Tibetan Plateau, with an average elevation of more than 4,000 m, is the world's largest plateau and Mount Everest (8,848 m), lying near the southern border of this plateau, is the world's highest peak.

Asia comprises more than 4 billion people living in 46 different States with a large range of differing cultures, environments, historical ties and government systems. Asia's wealth differs widely within and between States. In terms of nominal GDP, the largest economies in Asia are Japan (4,622.8 billion US$) and China (1931.7 billion US$) (UNDP, 2006). However, in terms of Purchasing Power Parity, the People's Republic of China has the largest economy (7,642.3 billion US$) in Asia and the second largest in the world while Japan (3,737.3 billion US$), India (3,389.7 billion US$) and South Korea (985.6 billion US$) are the world's third, fourth and eleventh largest economies respectively (UNDP 2006).

Remarkably, in most countries, central governments have played a key role in economic planning to achieve development goals, and in many instances, have also formulated environmental policies. Progress has been made in overall poverty reduction in many Asian countries. The number of people living on less than US$1 a day dropped by nearly 250 million from 1990 to 2001. The most severe problems are evident in South Asia, where almost half the children aged 0-5 years are malnourished. Sustained growth in China and the acceleration of India's economy contributed to such progress. Efforts at reducing malnutrition, however, have been less successful.

Notably, environmental protection is considered to be a fundamental element in achieving several Millenium Development Goals (MDGs) and is also a powerful engine for economic growth and poverty eradication.

South Asia has increased access for many to safe sources of drinking water, with India contributing substantially to the overall positive trend. Another positive sign is the significant progress made in improving energy efficiency and providing access to clean technology and fuels in Eastern Asia and Southern Asia, although energy efficiency continues to decline in Southeast Asia. The region still faces some formidable challenges in its effort to protect valuable natural resources and the environment. Land and ecosystems are being degraded, threatening to undermine food security. In addition, water and air quality are deteriorating while continued increases in consumption and associated waste have contributed to the exponential growth in the region's existing environmental problems. Furthermore, the region is highly subject to natural hazards. Notable examples include the 2004 Indian Ocean Tsunami, the 2005 Pakistan Earthquake, and the 2006 landslides in the Philippines.

There is evidence of prominent increases in the intensity and/or frequency of many of the extreme weather events such as heat waves, tropical cyclones, prolonged dry spells, intense rainfall, tornadoes, snow avalanches, thunderstorms, and severe dust storms in the region throughout the 1990s and beyond (IPCC-TAR). Impacts of such disasters range from hunger and susceptibility to disease, to loss of income and livelihoods, affecting the survival and human well-being of both present and future generations.

Indian continent in particular is facging strange climatic changes now—the arid/desert areas of Rajasthan are being flooded while other regions lack rainfall.

The environmental and social problems in Asia are not new; many of them are persistent and some are getting worse. The Brundtland report highlighted the risks twenty years ago, declaring that the loss of ecosystems, for example natural forests and wetlands, the degradation of water and air quality, and ever increasing consumption and associated waste generation have made people and ecosystems more vulnerable. Since 2000, Asia's GDP growth has surpassed that suggested by the Brundtland Commission, but ecosystems and human health continue to deteriorate. The region faces many years of challenges in attempting to preserve the environment while alleviating poverty and improving living standards with limited natural resources.

Remarkably Asia is the largest continent on Earth and spreads over four climatic zones, it has environmental and socio-economic problems common to the entire region, in sectors such as water, agriculture and land use, ecosystem and biodiversity, and population. The majority of the problems in these sectors are related to management issues and are exacerbated by the associated impacts of climate change.

Water Sector

Adequate water supply is one of the major challenges in Asia. The growing demand for water is driven by high population, increasing urbanization and the flourishing of economic and social development. Since Asian economies depend heavily on agriculture and hence on irrigation, this sector continues to be the most demanding on available water resources. Excessive withdrawals from surface waters and underground aquifers, pollution of freshwater resources by the industrial sectors and inefficient use of freshwater are also major causes of water stress.

Climate change has the potential to exacerbate water resource stresses in many countries of Asia (IPCC-TAR). The Arid characteristic of Middle Eastern countries makes water scarcity a major problem.

Extensive irrigation systems allow for agricultural production in this region, although this can increase salinity in groundwater aquifers. Combined with decrease in precipitation due to climate change, the Middle East faces a daunting task of ensuring fresh water supply for its sustainability.

For example, in Iran, agricultural water usage is about 81.4 billion cubic metres, which accounts for 94% of the total consumption. Drinking water accounts for about 5% of the overall consumption and the industrial sector for only 1% of the whole. The required water is primarily extracted from ground water resources, which supply 55% of water used. The rest is sourced from surface water using dams, irrigation and traditional methods (National Communication, Iran, 2003). Parts of Asia suffer from both water abundance as well as water scarcity.

It is significant to note that Southwest Bangladesh faces extreme water shortages as well as acute moisture stress during the dry months, adversely affecting both the ecology and agricultural production. Also, floods during the monsoon season inundate an average of 20.5% of the country and can flood as much as 70% (Mirza, 2002). The rate of retreat of glaciers in the Himalayan highlands and permafrost thawing in north Asia has rapidly increased in the last decade. The Himalayan glaciers feed seven of Asia's great rivers: Ganga, Indus, Brahmaputra, Salween, Mekong, Yangtze and Huang Ho, and ensure a year-round water supply to billions people. Human activities, such as land-use change, water storage, inter-basin transfers, and irrigation and drainage, are modifying the hydrological cycle in many river basins.

A change in the onset, continuity and withdrawal patterns of the summer monsoon in recent years has also led to considerable spatial and temporal variations in rainwater availability in parts of the continent (Lal, 2005). Furthermore, watershed quality over the past few years has been threatened by climatic variability and natural disasters. The impact extends to damage of sanitation facilities and contaminated groundwater (UNEP, 2005). There have been some attempts and actions to deal with water in Asia.

Major structural, technological, social and policy changes, including effective water-related policies, lifestyle changes and technological transfer, are needed to decrease water withdrawals and hence water stress in most Asian countries. In addition, improvements in water-use efficiency, especially in the irrigation sector, will have immediate positive impacts on water availability under warmer climate.

Pressures on Ecosystems

As a notable phenomenon, over the last two decades enormous pressures have been put on Asia's ecosystems to support the ever growing demand for natural resources. The most affected ecosystems are coastal and marine, forests and mountainous regions. Coastal ecosystems play an important role in the region. They are the locus of land-ocean interaction. The region has a long coastline and over half of Asia's more than 4 billion inhabitants live on or near the coast and depend directly on coastal resources, such as mangroves and coral reefs, for at least part of their livelihoods (Middleton, 1999).

Mangroves are vital to coastal ecosystems and fulfill multiple functions, such as providing forest products and habitat for fish and shellfish; protecting coastal land from erosion; enhancing biodiversity; and protection from inundation, for example the large mangrove forests in Phang Nga, Thailand, significantly mitigated the impact of the 2004 Indian Ocean Tsunami (UNEP, 2005).

The destruction and reduction of ecosystem services

and functions, in turn reduce their contribution to human well-being. Deforestation has caused the rapid reduction of timber production, especially of the timber only found in natural forests, thus affecting livelihoods (SEPA, 2004).

As per UNESCAP (2005), Heavy reliance on marine resources across the region has resulted in the degradation of many coral reefs, particularly those near major population centres. Moreover, the higher sea surface temperatures have led to severe bleaching of the corals in coastal regions of Asia. Approximately 60% of its coral reefs are estimated to be at risk and mining and destructive fishing practices are the greatest threats.

Sustainable Agriculture

Remarkably, Land degradation is a serious problem in Asia because of the threat to food security, human livelihoods and overall sustainability of the region. Problems include salinisation due to poor irrigation practices, soil erosion, desertification and overuse of water. Since agriculture is the main land use in Asia, land conservation as a tool of sustainable agriculture has been heavily emphasized.

It can promote rural development, as well as increase food security and ecosystem vitality. Immediate responses are reforestation, redefining protected areas and adopting integrated approaches such as Integrated Watershed Management. Good governance is the basic foundation of any land conservation strategy. Besides providing appropriate legal and policy mechanisms for administering land ownership, it can foster the active participation of civil society in land reform efforts and ensure equitable distribution of the benefits from agrarian development. Many farmers in Asia are women. Their contribution, however, tends to go unnoticed. Effective land management and conservation schemes take into consideration the need to recognize and protect the rights of female participants in agricultural areas and the benefits derived from those areas (FAO, 2003).

Significantly, The six geographical sub-regions of Asia primarily fall under the four climatic sub regions as categories by IPCC-TAR with it distinctive characteristics (Annex I, Table 2):

(a) *Boreal Asia:* It is located on the north of the continent between 50°N and the Arctic Circle. Boreal forests cover most of this region. Siberia, Russia, is the coldest region of the northern hemisphere in winter. Three Siberian rivers, the Ob, the Yenissey, and the Lena, contribute about 42% of total runoff from all rivers of the Arctic basin. Lake Baikal, in the heart of this region is one of the world's largest and oldest lakes and contains as much as 85% of Russian surface freshwaters.

(b) *Arid and semi-arid Asia:* It extends from 22°N to 50°N and from 30°E to 105°E and includes more than 20 countries of the Middle East and central Asia, many of which are landlocked. Many storms develop locally, and the maximum frequency of cyclones occurs in January and March, although very little precipitation is recorded over most of the region because of the lack of moisture. Grasslands, rangelands, and deserts dominate most of arid and semi-arid Asia.

(c) *Temperate Asia:* It extends from 22°N to 50°N and from 105°E to 150°E and includes eastern China, the Japanese islands, the Korean peninsula and Mongolia. Much of the natural forest in the region has long been destroyed. Broad plains have been cultivated and irrigated, and natural grasslands have been used for animal husbandry for centuries.

(d) *Tropical Asia:* Tropical region extends from 10°S to 28°N and from 50°E to 150°E and includes several countries of south Asia, which are influenced predominantly by the monsoons. The region is physiographically diverse and rich in biodiversity. Agriculture is the main industry in several countries. Exploitation of natural resources associated with rapid urbanization, industrialization, and economic development has led to increasing air and water pollution, land degradation, and other environmental problems in countries of this region.

Rainfall

Boreal Asia: There has been an observed average annual mean increase in surface air temperature of about 2.9°C in the past 100 years, which is more pronounced in the winter although summer temperatures in central Siberia have,

Observed changes in temperature and rainfall exhibited decreasing trends. Annual mean precipitation in Russia is decreasing about -4.1 mm/month/100 years, particularly during warm years. However, during the past 10–15 years, precipitation has increased, mostly during the summer-autumn period. Consequently, the levels of major aquifers have risen by 50–100 cm; the growth of groundwater storage has resulted in increasing ground river recharge and considerable low-water runoff.

Arid and semi-arid Asia: In most of the Middle East, mean annual surface temperature has risen by about 1.3°C during 1894--1997; in the arid regions of China, air temperature has increased since the 1970s. In Pakistan, annual mean surface temperature has a consistent rising trend since the beginning of 20th century. Annual mean rainfall is considerably low in most parts of the arid and semi-arid region and temporal variability is quite high.

Temperate Asia: The average annual mean surface temperature in Mongolia has increased by approximately 0.7°C over the past 50 years; consequently, noticeable changes have taken place in the length of the cold and warm seasons. Surface temperature in northeast China has increased in winter but decreased in summer; in southeast China surface temperature has decreased 1–2°C. In Japan, the surface air temperature has shown a warming trend during the past century. The East Asian monsoon greatly influences temporal and spatial variations in rainfall. Annual mean rainfall in Mongolia is 100–400 mm and is confined mainly to summer. Summer rainfall seems to have declined over the period 1970–1990 in Gobi; the number of days with relatively heavy rainfall events has dropped significantly. In China, annual precipitation has been decreasing continuously since 1965. The summer monsoon is reported to be stronger in northern China during globally warmer years although drier conditions have prevailed during globally colder years.

Tropical Asia: Several countries have reported increasing surface temperatures. In Vietnam, annual mean surface temperature has increased approximately 0.32°C over the past three decades and in Sri Lanka a gradually increasing trend is suggested of about 0.30°C per 100 years. The warming trend over India has been reported to be about 0.57°C per 100 years. Hills and mountain ranges cause striking spatial variations in rainfall. In India, long-term time series of summer monsoon rainfall have no discernible trends, but decadal departures are found above and below the long time average alternatively for three consecutive decades. Recent decades have exhibited an increase in extreme rainfall events over northwest India during the summer monsoon and the number of rainy days during the monsoon along the east coast has declined. 27. Extreme weather events such as heat waves associated with extreme temperatures, extra-tropical and tropical cyclones, prolonged dry spells, intense rainfall, tornadoes, snow avalanches, thunderstorms, and dust storms are known to cause adverse effects in widely separated areas of Asia.

There is some evidence of increases in the intensity or frequency of some of these extreme weather events on regional scales throughout the 20th century, although data analyses are relatively poor and not comprehensive (Balling and Idso, 1990; Bouchard, 1990; Agee, 1991; Yu and Neil, 1991; Ostby, 1993; Bardin, 1994; Born, 1996). There are also reports of an increase in thunderstorms over the land regions of tropical Asia (Karl et al., 1995).

The frequency and severity of wildfires in grasslands and rangelands in arid and semi-arid Asia have increased in recent decades (IPCC-TAR). Permanent glaciers in mountainous areas have vacated large areas during the past few decades, resulting in increases in glacial runoff and Glacier Lake Outburst Floods (GLOFs) and an increased frequency of events such as floods, mudflows and avalanches affecting human

3. *Extreme Events and Severe Weather* Systems settlements. However, as mountain glaciers continue to disappear over the next decades, the volume of summer runoff will be reduced as a result of loss of ice resources with serious consequences for downstream agriculture. For example, low- and mid-lying parts of Central Asia are likely to change gradually into more arid, interior deserts,

and lower Himalayan Countries will have more intense and frequent drought.

Recently, years, GLOFs have happened in Nepal, India, Pakistan, Bhutan, and China in the Himalayan region. The most significant GLOF event since detailed recording started occurred in Dig Tsho Lake at Bhote Kosi in Nepal in 1985. It destroyed a hydropower project, numerous foot-bridges, and trekking trails and caused loss of many lives. Millions of hectares of protected areas and many human settlements are located in the downstream of potentially dangerous glacial lakes. GLOF events have widespread impacts on the region's socio-economic systems, hydrology, and ecosystems. Temperate Asia has experienced many droughts in the 20th century.

In China, droughts in 1972, 1978, and 1997 have been recorded as the most serious and extensive. In Japan, drought disasters are significantly more frequent during years following ENSO warm events than in normal years. Floods are also a threat, for example a large number of severe floods have occurred in China, predominately over the middle and lower basins of the Yangtze (Changjiang), Huanghe, Huaihe, and Haihe Rivers (Ji et al., 1993) and severe flooding with daily rainfall exceeding 25 cm struck during July and August 1998 in South Korea.

In Tropical Asia, droughts, floods and cyclones are the key natural disasters. In India, chronically drought-affected areas cover the western parts of Rajasthan and the Kutch region of Gujarat. But since 2006, Barmer and Jodhpur regions are getting heavy rainfall causing rever flood conditions. Drought conditions have also been reported in Bihar and Orissa States.

In Bangladesh, about 2.7 Mha are vulnerable to drought annually (Mirza, 1998). Drought or near-drought conditions also occur in parts of Nepal, Papua New Guinea, and Indonesia, especially during El Niño years. In India, Laos, the Philippines, and Vietnam, drought disasters are more frequent during years following ENSO events. At least half of the severe failures of the Indian summer monsoon since 1871 have occurred during El Niño years (Webster et al., 1998). Floods are also common in this region. For example in Bangladesh the total flood-prone area is 3.1 Mha and in India it is about 40 Mha (Mirza and Ericksen, 1996). In the event of enhanced anomalous warming of the Eastern Equatorial Pacific Ocean, such as that observed during the 1998 El Niño, a higher frequency of intense extreme events across Asia is possible.

In fact, there is general agreement among global climate models (GCMs) of a significant acceleration of warming in the 21st century in Asia over that observed in the 20th century. Predictions indicate that higher latitudes in Asia will warm more than the global average and will receive disproportionately more precipitation, while mid-continental mid-latitude areas may become drier, depending on the effects of aerosols (IPCC-TAR; Gao et al., 2003).

The area-averaged annual mean warming is projected to be about 3°C by the 2050s and about 5°C by the 2080s over the land regions of Asia as a result of future increases in atmospheric concentration of greenhouse gases. The rise in surface air temperature is projected to be most pronounced over North Asia in all seasons. In general, projected warming over all sub-regions of Asia is higher during northern hemispheric winter than during summer for all time periods.

An increase in annual precipitation in most of Asia

during the 21st century is predicted with the relative increase being largest and most consistent in North and East Asia (Lal, 2003; Kwon et al., 2004; Boo et al., 2005; Japan Meteorological Agency, 2005; Kurihara et al., 2005). There is a predicted decline in summer precipitation over the central parts of arid and semi-arid Asia, leading to expansion of deserts and periodic severe water stress conditions. Increased rainfall intensity, particularly during the summer monsoon, could increase flood-prone areas in East Asia, South Asia and Southeast Asia.

Future Projects

While there is no evidence that tropical cyclone frequency may change (Trenberth, 2005), an increase of 10-20% in tropical cyclone intensities (including the near-storm precipitation rates and destructive potential) for a rise in sea surface temperature of 2 to 4oC relative to the current threshold temperature is likewise projected in East Asia, Southeast Asia, and South Asia (Knutson and Tuleya, 2004; Emanuel, 2005). Amplification in storm surge heights could result from the occurrence of stronger winds with increase in sea surface temperatures and low pressures associated with tropical storms resulting in an enhanced risk of coastal disasters along the coastal regions of East, South and Southeast Asian countries. The impacts of increase in cyclone intensities in any location will be determined by any shift in the cyclone tracks.

In fact, global warming has accelerated melting of glaciers, with the vast Himalayan glaciers showing the fastest rate of retreat. Recent studies by the International Commission of Snow and Ice (ICSI) and the DFID, have predicted that Himalayan glaciers might disappear within 40 years time causing water stress to millions in India and China depending on glacier fed rivers. The Chinese Academy of Sciences states that as many as 64 % of China's glaciers will not exist by 2050 if current trends continue unabated.

Future changes in sea level rise are predicted to be geographically variable. An additional 0.5m sea level rise is projected for the Arctic during this century (ACIA, 2004). The rising rates of sea level vary considerably from 1.5 to 4.4 mm per annum along the East Asia coast due to regional variation in land surface movement (Mimura et al., 2004). The projected rise of mean high water level is greater than that of mean sea level (Chen, 1991; Zhang et al., 2000). Thus projected relative sea level rise (RSLR), including that due to thermal expansion, tectonic movement, ground subsidence and the trend of rising river water level, is 40-60cm, 50-70 cm and 70-90 cm in the Zhujiang, Changjiang and Huanghe Deltas, respectively by the year 2050 (Li et al., 2004a, b).

National Programmes for Adaptation and Climate Change

Almost all Asian Counties have submitted their NC to the UNFCCC to comply with their commitments under the Convention. Three Asian countries, Bangladesh, Bhutan and Cambodia, have also submitted their National Adaptation Programmes of Action (NAPAs). Other LDCs in Asia are finalizing their NAPAs. The levels of assessment of impacts, vulnerabilities and adaptation to climate change and reporting varies widely by region and countries due to differences in institutional capacity and economic strengths. Some countries have reported that adaptation measures have already been adopted to address climate change related problems.

For example, the Law on Water Resources of the People's Republic of China has been promulgated to safeguard and normalize the management of water resources. The Project of South-to-North Water Diversion has been launched in order to increase water supply to Beijing, Tianjin and the Shandong Peninsula. The construction of upgrading projects for water-saving in 226 large irrigated areas and 200 demonstration projects for water-saving irrigation were launched to raise water use efficiency (Initial National Communication on Climate Change, China, 2004). Terracing of slopes, one of the most effective amelioration activities, has been practiced in many mountainous and low mountain regions for a long time, for example Azerbaijan, Nepal, Indonesia and Bhutan. Besides prevention of erosion processes, terracing of slopes enables additional areas to be cultivated and improves land productivity. The NC of Azerbaijan reported that productivity of terraced slopes would increase by 8-10 times once the cultivation of fodder crops is practiced. Increase of terraced slopes' productivity is maintained through the improvement of plant water regime as a result of 2-3 fold decrease of surface drain.

Many NCs give an emphasis on "no-regret" adaptation options, particularly for the water and agriculture sectors, in order to meet national development goals and promote sustainable development (National Communication of India, Saudi Arabia, Bangladesh, China etc). For example, China stated that they are promoting sustainable development and utilization of water resources; reinforcing adaptive capacity of the water resource system; and thus reducing the vulnerability of the water resource system to climate change. Iran suggests a number of methods for conserving water used in agriculture such as improving irrigation efficiency by lining existing earth irrigation canals, using pipes as water conduits and the reuse of drainage water. The government of Saudi Arabia has implemented measures to protect the sustainability of water resources. These measures include execution of a long-term investigation programme to assess the availability of groundwater and surface water resources in different regions and supporting the drilling of thousands of wells, under the supervision of the Ministry of Agriculture, for domestic, agricultural, and industrial purposes.

Collaboration

Considering the importance and need for capacity-building to deal with climate change, a number of networks are working in the Asia Region. Most of them are supporting research, generating knowledge, bringing new ideas and debate, and sharing knowledge among different climate communities including researchers, policy makers and negotiators. The Asia-Pacific Network for Global Change Research (APN) is an inter-governmental network whose primary purposes are to facilitate global environmental change research in the Asia-Pacific region, increase developing country participation in that research, and strengthen links between the science community and policy makers. It promotes, encourages and supports research activities on long-term global changes in climate, ocean and terrestrial systems, and on related physical, chemical, biological and socio-economic processes.

It realized that global change research cannot succeed without a high level of international cooperation and thus supports collaborative projects implemented under international research programmes. Projects being implemented by APN can be categorized as Capacity-building, Comprehensive Research, and Adaptation to Climate Change, for example the 'Linking Climate Change Adaptation to Sustainable Development in Southeast Asia' project.

In Asian LDCs, the network has built capacity through transferring tools and methodologies, assessment, training and awareness raising. It has also generated new knowledge on climate change and its impacts on water resources and agriculture.

Remarkably, START (SysTem for Analysis, Research and Training), provides an international framework for capacity-building in scientific knowledge and technology to conduct regional and local research and to inform and influence decision-makers. START has forged strong linkages with regional governmental organizations, such as the Asia-Pacific Network for Global Change Research (APN), the Inter-American Institute (IAI), the New Partnership for Africa's Development, and the European Commission (EC). START furthermore collaborates with other organizations dedicated to building scientific research capacity across disciplines in developing regions, such as the Third World Academy of Science (TWAS), the International Foundation for Science (IFS), and the United Nations University (UNU). The START Regional Networks include Southeast Asia, South Asia and Temperate East Asia.

It is significant START completed its Assessments of Impacts and Adaptation to Climate Change (AIACC) project in Asia where potential impacts of climate change and vulnerability and adaptation assessment have been carried out for some sectors and countries. Advancing Capacity to Support Climate Change Adaptation (ACCCA) is an ongoing project of START focusing on community-based adaptation measures to weather related disasters, integration of adaptation strategies into development policies by effectively communicating climate risks and adaptation measures in agriculture and water sectors and mainstreaming climate change adaptation in watershed management and upland farming.

The Centre for International Forest Research (CIFOR) is an international research and global knowledge institution committed to conserving forests and improving the livelihoods of people in the tropics. CIFOR's high impact research helps local communities and small farmers gain their rightful share of forest resources, while increasing the production and value of forest products. CIFOR's three research programmes address the needs of the rural poor as well as environmental concerns.

The Environmental Services and Sustainable Use of Forests Programme oversees research on biodiversity, carbon, fires, watershed functions, and the sustainable management and harvesting of forest products; the Governance Programme examines the process of making and implementing decisions about the management of forests by people and organizations beyond the scale of the individual household or small enterprise; and the Forests and Livelihoods Programme closely investigates how forest resources and their management, use and trade contribute to the livelihoods of the rural and urban poor.

In the similar veinInternational Center for Integrated Mountain Development (ICIMOD), with financial support from UNEP, is assisting governments with the development of an inventory of glacial lakes; establishment of early warning systems; and development of adaptation measures. Inventories have been completed for Bhutan, Nepal, Astor basin in Pakistan, Pumqu basin in China, and Tista basin in India. Twenty glacial lakes in Nepal and 24 in Bhutan were identified as potentially dangerous and were recommended for further detailed investigation.

The International Water Management Institute (IWMI) is a non-profit scientific organization funded by the Consultative Group on International Agricultural Research (CGIAR). IWMI's research agenda is organized around four priority themes covering key issues relating to land, water, livelihoods, health and environment. The Institute concentrates on water and related land management challenges faced by poor rural communities. The challenges are those that affect their nutrition, livelihoods and health, as well as the integrity of environmental services on which these depend. IWMI works through collaborative research with partners in the North and South, to develop tools and practices to help developing countries eradicate poverty and better manage their water and land resources. For disaster preparedness, rapid response and assessment, a Remote Sensed and GIS Based Knowledge Base System for Sri Lanka (KBS-LANKA) is being implemented by IWMI.

The Institute for Global Environmental Strategies (IGES) is a research institute that conducts pragmatic and innovative strategic policy research to support sustainable development in the Asia-Pacific region. IGES collaborates with a broad range of stakeholders, such as national governments, non-governmental organisations, businesses and citizens groups, to carry out research, aiming to ensure that the results are reflected in the policy-making process. It has produced a number of policy documents on adaptation with special focus on adaptation beyond 2012.

South-South-North (SSN) is a network-based non-profit organisation working in the fields of climate change and social development. They aim to reduce poverty by building capacity in the South and delivering community based mitigation and adaptation projects. SSN has developed a methodology for preparing community based adaptation projects known as SouthSouthNorth Adaptation Project Protocol (SSNAPP).

The World Conservation Union (IUCN) is the world's largest and most important conservation network working on climate change issues in Asia. This organization is involved in the National Capacity Self-Assessment Programme.

The Asian Disaster Preparedness Center (ADPC) is a non-profit organization supporting the advancement of safer communities and sustainable development, through implementing programmes and projects that reduce the impact of disasters upon countries and communities in Asia and the Pacific. ADPC is working on raising awareness, strengthening institutions, providing training and mainstreaming disaster reduction in development management in Asia.

The Red Cross/Red Crescent Climate Centre aims to support the International Federation of Red Cross and Red Crescent Societies (IFRC) and other interested parties to reduce people's vulnerability to climate risks. The activities of the Climate Centre include: (a) providing education and communication support about climate risk management within the Red Cross and Red Crescent Movement and among the general public; (b) Supporting the integration of climate risk management into Red Cross/Red Crescent programmes through the ʹPreparedness for climate change' programme that formally started on January 1, 2006; (c) Bringing concerns about the impacts of climate change on vulnerable people and experiences with climate risk reduction to national and international policy makers; (d) Analyzing, documenting and sharing knowledge and experiences on climate risk management. For example, the climate change group is working in Indonesia on integrated community based risk reduction and in Vietnam on community-based climate adaptation.

Capacity Strengthening of LDCs for Adaptation to Climate Change (CLACC) is an initiative of southern institutes working on sustainable development. Initiated in 2004 by the International Institute for Environment and Development (IIED) and the Regional and International Networking Group (RING) partner institutions, CLACC receives financial and technical support from a number of northern institutes. CLACC aims to strengthen the capacity of organizations working with civil society in environment and development while integrating climate change issues in the planning and implementation of projects.

The Energy and Research Institute (TERI) in India develops solutions to global problems in the fields of energy, environment and current patterns of development which are largely unsustainable. The central element of TERI's philosophy has been its reliance on entrepreneurial skills to create benefits for society through the development and dissemination of intellectual property. The institute not only identifies and articulates intellectual challenges straddling a number of disciplines of knowledge but also mounts research, training and demonstration projects leading to the development of specific problem-based advanced technologies that help carry benefits to society at large. TERI works closely with several multilateral organizations, national governments, corporate organizations, NGOs, and academia around the world.

The Bangladesh Centre for Advanced Studies (BCAS) is an independent, non-profit, non-governmental, policy, research, and implementation institute working on sustainable development at local, national, regional and global levels. BCAS envisions promoting people-centred sustainable development by applying and advancing scientific, technical and local knowledge through research, by developing models, demonstration, policy advocacy and project implementation. It works with leading NGOs, research institutes, universities and government agencies in South Asia and other parts of the world.

Asia still faces formidable challenges in its effort to protect valuable natural resources and the environment. Land and ecosystéms are being degraded, threatening to undermine food security. In addition, water quality and availability are deteriorating while continued increases in consumption and associated waste have contributed to the exponential growth in the region's existing environmental problems. Furthermore, the region is highly subject to natural hazards.

Countries in the Asian region have assessed impacts, vulnerability and adaptation to climate change using model-based and vulnerability-based approaches. The vulnerability-based or bottom-up approach was found to be good for identification of adaptation measures which are more people centred. The availability of data with the necessary geographic detail is the major limitation rather than computational capability or basic understanding particularly for agriculture, water and coastal zones. Climate change induced key biophysical vulnerabilities vary widely across the region due to differences in physical, social and economic circumstances.

It is revealed from the NCs and NAPAs that the most vulnerable sectors are agriculture and food security, water resources, natural ecosystems, forestry and biodiversity, coastal zones and sea level rise, natural disasters and human health. The projected impacts of climate change are likely to exacerbate water stress and shortages in some regions and increase the frequency and intensity of floods and droughts. A wide range of precautionary measures are available at the regional and national level to reduce economic and social impacts of natural disasters. Integration of disaster risk reduction and climate change is emerging in many Asian countries. Adaptive capacities vary between countries, depending on social structure, culture, economic capacity, and the level of environmental disruptions. Limiting factors include poor resource and infrastructure bases, poverty and disparities in income, weak institutions, and limited technology. The challenge in Asia lies in identifying opportunities to facilitate sustainable development with strategies that make climate-sensitive sectors resilient to climate variability.

Generally, financial, technological and institutional barriers usually hamper the implementation of adaptation measures to climate variability and change in many Asian countries, particularly LDCs. For example, the current water policy of India aims at integrated water resources development and management to tackle water stress, but its implementation is constrained by financial and technological limitations.

Many of the problems associated with climate impacts are also priorities on the development agenda and have been for a long time, irrespective of climate change. Environmental degradation and links between the environment and development have been well documented and analysed since the 1970s. Enough is known of the risks posed by climate variability and climate change to develop and implement adaptation projects immediately. Specific focus is necessary on the following issues:

1. Adaptation should aim to strengthen traditional coping mechanisms: optimizing current systems whilst building flexibility to cope with the uncertainties posed by climate change. Introducing new technology can be sustainable where it strengthens and builds on traditional approaches and reinforces local knowledge;
2. Development of effective adaptation strategies should take a more systems-oriented, people centred approach, emphasizing multiple interactive stresses, with less dependence on climate scenarios;
3. Implementation of adaptation measures identified in National Communications and National Adaptation Programmes of Action should be supported effectively;
4. An assessment of exiting coping strategies and its effectiveness under the warmer climate is necessary to modify practices for readjusting with changing condition that climate change will bring;
5. Capacity-building and enhancement of technical capacity is needed to assess, plan and integrate adaptation needs into sectoral development plans that deal with the adverse impacts of climate change;
6. Notably, the development of educational and public awareness programmes on climate change, the provision of public access to information on

climate change issues and public participation (including NGOs) are important components of both the implementation of the UNFCCC and the development of national action plans;

7. There is a need to support mainstreaming of adaptation into sectoral policy, particularly water, agriculture, coastal zones and managing natural ecosystems;
8. More work is needed to promote mechanisms to increase funds for adaptation to climate change.

In order to address adaptation to climate change, including variability and extreme events, it is vital to continue to research the circumstances of the poor and vulnerable, both *for* the poor and vulnerable and *on* the poor and vulnerable. This requires stakeholder involvement, and will not be based exclusively on a top-down approach. This research must be communicated in a clear, targeted manner to relevant stakeholders including the communities themselves, NGOs active in the communities, local and national authorities, higher education institutes, global scientists, policymakers and decision makers.

5

CLIMATE CHANGE: AWARENESS & INTERNATIONAL ACTION

> We work day after day, not to finish things; but to make the future better . . . because we will spend the rest of our lives there.
>
> —*Charles F. Kettering*

Global Warming Awareness

It is strange, some twenty-five years ago if you made a trip to the local library and perused the periodical section for articles on global warming you'd probably have come up with only a few abstracts from hard core science journals or maybe a blurb in some esoteric geopolitical magazine. As an Internet search on global warming now attests, the subject has become as rooted in our public consciousness as microwave cooking or something popular like that. Perhaps all this attention is deserved. With the possible exception of another world war, a giant asteroid, or an incurable plague, global warming may be the single largest threat to our planet. For decades human factories and cars have emitted billions of tons of greenhouse gases into the atmosphere, and the climate has begun to show some signs of warming. Many see this as a harbinger of what is to come.

If we don't curb our greenhouse emissions, then low-lying nations could be awash in seawater, rain and drought patterns across the world could change, hurricanes could become more frequent, and El Niños could become more intense. On the other hand, there are those, some of whom are scientists, who believe that global warming will result in little more than warmer winters and denser vegetation. They point to the flaws in scientists' measurements, the complexity of the climate, and the uncertainty in the climate models used to predict climate change. They claim that attempting to lower greenhouse emissions may do more damage to the world economy and human society than any amount of global warming.

In truth, the future probably fits somewhere between these two scenarios. But to gain an understanding of global warming, it is necessary to get to know the science behind the issue. Our Warming Planet or for that matter any substantial warming of the Earth's surface, begins with the sun. Except for relatively small fluctuations due to sunspot activity, the amount of radiation from the sun that reaches the Earth has been fairly constant from year to year and century to century. If you were to travel to the outer reaches of the atmosphere and hold up a flat surface perpendicular to the sun's rays for several years during the daylight hours, you'd find that about 1,368 Watts per square meter on average would hit that surface. Not all of that energy would be absorbed by the Earth. Roughly 30 percent of the total solar energy that strikes the Earth is reflected back into space by clouds, atmospheric aerosols, reflective ground surfaces, and even ocean surf. The remaining 70 percent is absorbed by the land, air, and the oceans. The absorbed light is mostly in the form of ultraviolet, visible, and near-infrared solar radiation. Absorption of solar energy heats up our planet's surface and atmosphere and makes life on Earth possible. The energy does not stay bound up in the Earth's environment forever. If it did, then the Earth would grow hotter and hotter until its temperature exceeded that of the sun. Instead, as the rocks, the air, and the sea heat, they emit thermal radiation.

Much of this thermal radiation, which is largely in the form of long wave infrared energy, travels directly out into space, leaving the Earth and allowing it to cool. Such radiation is invisible to our eyes, but our hands can feel it radiating from a fire or a car engine. Some of this outgoing long wave infrared radiation, however, is re-absorbed by water vapor, carbon dioxide, and other greenhouse gases in the atmosphere and is then re-radiated back toward the Earth's surface. On the whole this re-absorption process is good. If there were no greenhouse gases or clouds in the atmosphere, the Earth's average surface temperature would be a very chilly -18°C instead of the comfortable 15°C that it is today. What has many people worried now is that over the past 250 years humans have been artificially raising the concentration of greenhouse gases in the atmosphere. Our factories, power plants, and cars and other vehicles burn coal and gasoline and spit out a seemingly endless stream of carbon dioxide. We produce millions of pounds of methane by allowing our trash to decompose inland fills and by breeding large herds of methane-belching cattle. Nitrogen-based fertilizers, which we use on nearly all our crops, release unnatural amounts of nitrogen oxide into the atmosphere. Once these carbon-based greenhouse gases get into the atmosphere, they stay there for decades or longer. According to the Intergovernmental Panel on Climate Change (IPCC), since the industrial revolution, carbon dioxide levels have increased 31 percent and methane levels have increased 151 percent. Pale climate readings taken from ice cores and fossil records have shown that these gases, two of the most abundant greenhouse gases, are at their highest levels in the past 420,000 years. Many fear that the increased concentrations of greenhouse gases have prevented additional thermal radiation from leaving the Earth. In essence, these gases are trapping excess heat in the Earth's atmosphere in much the same way that a windshield traps solar energy that enters a car. Much of the available climate data appears to back these fears. Temperature data gathered from many different sources all nearly always coincide with large increases in Earth surface temperatures.

Conversely, ice ages are almost always accompanied by a decrease in carbon dioxide. Logic dictates that, as third world nations develop their economies and first world nations consume more energy, greenhouse gas concentrations will continue to rise. Though scientists have not reached a consensus, most leading researchers and organizations purport that the average surface temperature of the Earth will increase along with increasing emissions. According to the IPCC, the surface temperature could rise by between 1.4°C and 5.8°C by the end of the this century. Scientists at Goddard Institute for Space Studies, NASA's division spearheading climate modeling efforts, report that we should expect between 0.5°C and 1°C over the next fifty years.

At first glance, spherically, these numbers probably do not seem threatening. After all, temperatures typically change a few degrees whenever a storm front moves through. Such temperature changes, however, represent day-to-day regional fluctuations. When surface temperatures are averaged over the entire globe for extended periods of time, it turns out that the average is remarkably stable.

Rarely in the Earth's history has the average surface

temperature changed as dramatically as the changes that scientists are predicting for the next century. During the last ice age some 20,000 years ago, for instance, the Earth was roughly 5°C cooler than it is today. Since then it has warmed up, although not steadily, to present levels. That's an increase of roughly 1°C every 4,000 years. Current global warming awareness scenarios predict this bare minimum increase over the next century.

How all of this warming will alter the weather is more uncertain. It's much easier for scientists to forecast the Earth's average surface temperature than it is to forecast how much rain will fall in, say, Boise, Idaho, during the next in 50 years. So far scientists have not been able to pinpoint with certainty any changes in weather due to global warming over the last century. Most of that "weird" weather we've been experiencing that unusually warm fall or that particularly wet winter is due to normal, regional changes in the weather.

Some scientists believe that global warming will continue to have relatively little impact on the day-to-day climate conditions. Others purport that future changes will likely be subtle, and they will spread over large areas of the globe from decade to decade and creep up on us like old age. Still others hypothesize that when the Earth's surface temperature reaches some critical threshold, the heat will trigger relatively drastic changes to the atmosphere and the oceans and transform the Earth's weather patterns in a matter of years. Not surprisingly, many scientists speculate that such changes in the climate will probably resulting more hot days and fewer cool days. According to the IPCC, land surface areas will increase in temperature over the summer months much more than the ocean. The mid-latitude to high latitude regions in the Northern Hemisphere areas such as the Continental United States, Canada, and Siberia will likely warm the most. These regions could exceed mean global warming by as much as 40 per cent. Forecasts for precipitation and weather are cloudier. Right now the IPCC reports that the amount of precipitation, especially in the mid-latitude to high-latitude regions of the Northern Hemisphere, will likely increase.

They believe, however, that it will come in the form of bigger, wetter storms, rather than in the form of more rainy days. So it's more probable that the increase in rain will only serve to tax our drainage systems rather than benefit vegetation or replenish natural, underground aquifers.

As to larger more destructive weather patterns, hurricanes will likely increase in intensity due to warmer ocean surface temperatures. And researchers speculate that El Nina events may increase in intensity for the same reason. The outlook for rising sea levels is nothing like the deluge portrayed in fantasy films.

According to the IPCC, over the next century, sea levels are likely to rise between 0.09 and 0.88 meters. The rise will mainly be due to seawater expanding from the increased ocean temperatures and run-off from the melting of continental glaciers and a slight melting of the Greenland Ice Sheet. For now, the West Antarctic Ice Sheet, which could raise our sea levels dramatically, will probably stay in place. It may even gain more mass due to an increase in precipitation over the next century. But, if somehow the entire Greenland Ice Sheet melted and the West Antarctic Ice Sheet fell into the sea, the sea level would rise roughly 10 meters. This is probably impossible over the next century, but there is the danger that global warming could initiate ice sheet changes that will continue to develop over future centuries. Should global warming continue, many biologists envision the alteration of natural habitats? Some of this change maybe for the better. Higher levels of CO_2 and warmer temperatures may cause forests to become more lush and vigorous. Warmer ocean waters on the open ocean could be beneficial to fish and algae on the high seas. Unfortunately, most changes will likely be for the worst. Plants and animals in mid-latitude regions, such as nut-bearing oaks in the Midwestern United States, may find themselves in warmer environments where they cannot survive. Rising sea levels may inundate delicate coastal wetlands with brackish waters, which could drive out certain types of fish and kill wetland vegetation. Warmer ocean temperatures around the coast could overheat many types of coral, killing them and many of the animals that depend on them.

Now coming to the human health, those hit hardest will probably be residents of poorer countries that do not have the funds to fend against changes in climate. A slight increase in heat and rain in equatorial regions would likely spark an increase in vector-borne diseases such as malaria. More intense rains and hurricanes could cause more severe flooding anymore deaths in coastal regions and along riverbeds. Even moderate rise in sea level could threaten the coastlines of low lying islands such as the Maldives. All across the globe, hotter summers could lead to more cases of heat stroke and deaths among those who are vulnerable, such as older people with heart problems. The warmer temperatures may also lead to higher levels of near-surface ozone from cars and factories, which would likely cause more code red air quality days and hospital admissions for those with respiratory problems.

The severity of these environmental changes will be largely dependent on how much the Earth's surface warms over the next century. As the wide range of estimates for average global surface temperature suggests, researchers haven't exactly reached a consensus. The reason for the wide range simply comes down to the difficulty inherent in predicting the outcomes of current trends in both human society and the Earth's climate system. To get their estimates for future warming, scientists must first discern how much human industry and expansion will impact the Earth over the next 100 years. Researchers typically reviews wide range of socioeconomic data such as worldwide population trends and then come up with varying scenarios describing mankind's future. For the IPCC report, 16 such scenarios were developed. The worst case scenarios, typically known as the "business as usual" scenarios, go on the assumption that population growth in developing nations will proceed unabated and the entire world will continue to use more and more fossil fuel per capita. The best case scenarios envision a world in which environmentally friendly technologies such as fuel cells and solar panels replace much of today's fossil fuel combustion and the population increase is halted by an improved standard of living worldwide. Using these projections, researchers arrive at estimates for how much carbon dioxide, soot, ozone and other pollutants people will put into the air over the next 100 years. Obviously, the "business as usual" scenarios have people producing the most pollution, and the eco-friendly scenarios have people producing the least pollution. But projections of greenhouse gas concentrations alone cannot tell scientists how much the Earth's surface will warm or the climate will change. To make forecasts, they must employ climate models, which are essentially computer simulations of the climate. These models are a bit like those computer programs detectives use to envision what missing persons would look like ten years after their disappearance.

Some of these models simply produce forecasts of the Earth's temperature, while others are built to predict other changes in the Earth's atmosphere and oceans. Even the simplest of these models can be exceedingly complex. When piecing together models to estimate the Earth's average surface temperature, scientists must take into account everything on the Earth that blocks outgoing thermal radiation or reflects sunlight into space as well as possible changes in the radiation emitted by the sun itself. There are, in fact, amyriad of unnatural and natural factors in addition to greenhouse gases that could sway global surface temperatures one way or the other in the future. Some of these influences are human made and some are natural. Some directly impact the amount of radiation absorbed by the Earth, and scientists expect others will be triggered as the climate heats up.

One example of a cooling influence that may counter the greenhouse effect is sulfate aerosols. When fossil fuels burn, they not only release greenhouse gases, but also sulfur dioxide. The sulfur dioxide gets into the air and mixes with oxygen to create sulfate aerosol particles, which reflect sunlight. Though these aerosols don't stay in the air as long as most greenhouse gases, they may cool regions of the world where their emissions are heavy, such as the eastern United States, China, and the Indian sub-continent. Changes in clouds due to the warming of the atmosphere, on the other hand, could greatly add to the warming effect. As the Earth heats up, the cloud composition in the atmosphere could change dramatically. Low lying clouds could evaporate during the daytime hours and more high flying cirrus clouds could form as surface heat causes the air to rise. Since low lying clouds tend to reflect sunlight and higher clouds absorb heat radiation, the overall effect of these changes would be an increase in the amount of energy trapped in the atmosphere. On the other hand, the response of clouds to global warming is so uncertain that changes in clouds might act to reduce, not increase, the amount of surface warming over the next century. As if these warming and cooling influences weren't enough to keep track of, some evidence has shown global warming may have a delay built into it. Given the level of carbon dioxide in the atmosphere now, a number of Earth scientists calculate that the Earth should be 0.5°C warmer than it is today. Temperature readings obtained from several hundred meters below the ocean's surface suggest that this extra energy could be lurking down there. The theory is that as the Earth has warmed, much of the excess energy has gone into heating the upper layers of the ocean, giving rise to a temperature imbalance between the surface waters and the deep waters. It is thought that convection currents in the ocean may have transported some of this excess heat in the surface waters down deep, effectively removing it from the surface of our planet. Were this process to continue, the temperatures of the lower layers of the ocean would eventually increase until they are once again in balance with the surface waters. At this point, the excess heat in the upper layers would no longer be drawdown, and the Earth would warm to a higher level. So even if we drastically lower our emissions today, we could still be informing a 0.5°C additional warming. Climate modelers must consider dozens of such factors, boil them down into equation form, and pack them into their models. Not all models are built alike. There is quite a lot of disagreement among Earth scientists as to how much of a role factors such as aerosols and clouds will play in heating the Earth and how they should be incorporated into the models.

For instance, NASA climate modelers at GISS have evidence that black carbon aerosol particles (soot) contributes significantly to warming of the lower atmosphere, since they absorb incoming radiation. The IPCC, on the other hand, estimates that black soot plays only a very small role in warming. Each research group or agency builds its models accordingly, and the choices made influence the forecasts derived from the models.

Even when modelers do agree on the mechanisms involved, many of these factors have a great deal of uncertainty associated with them. While the general consensus among scientists is that global warming is real and its overall effects are detrimental, there're still some prominent scientists who feel that the threat of global warming has been greatly exaggerated. Skeptics take issue with the basic temperature data that demonstrate the Earth's temperature has increased over the last century. Most of the pre-satellite, pre-1970 data were collected in urban areas using many types of thermometers that were spread far apart. Such measurements are subject to human error and do not give a clear depiction of ocean temperatures. Until satellite data are collected for several more decades, some researchers feel that the temperature data remain too unreliable to take at face value.

Another point of contention is that no one has ever proven outside of the laboratory whether global warming occurs as result of carbon dioxide. Scientists have ample fossil evidence that shows that carbon dioxide levels in the atmosphere have risen as the Earth grows warmer, but no one has yet shown that a rise in carbon is responsible for the past temperature increases. It is still possible that the warming in the distant past could have triggered the rise in carbon dioxide. As far as forecasts of future warming are concerned, skeptics point to the uncertainties inherent in the models researchers are using. There are a couple of dozen models currently in use that forecast everything from the average surface warming of the planet to complex interactions global warming will have with the Earth's atmosphere and weather systems. As mentioned, each of these models can generate a different answer depending on projections for future human emissions, the uncertainty in how the climate will respond, and what scientists decide to include in the models. Many feel there is still too much we do not understand about the climate or human society to take stock in any forecasts as of yet. But perhaps the one aspect of global warming research that nearly everyone agrees on is that more has to be done. NASA's Earth Science Enterprise is involved in a number of projects that aim to monitor and analyze climate change. A number of Earth Observing System (EOS) satellite missions have either already been launched or are stated to be launched over the next several years.

Instruments aboard these satellites will take unprecedented measurements of the Earth and the sun that are relevant to climate change. The EOS Terra satellite, which was launched in 1999, retrieves global readings of land surface temperatures, snow cover, atmospheric aerosol levels, cloud properties, methane, vegetation density, and a host of other variables that influence climate on a global scale. The measurements are taken over the entire globe once a day, often taken at resolutions of 1 km or less. Taking measurements by satellite is much more efficient, consistent, and timely than taking measurements in situ on the planet's surface. Since the early 1970s, researchers at GISS have been constructing climate models with the data and knowledge gathered in part from satellite readings. They've constructed a number of radioactive forcing models to try to understand the manner in which the Earth absorbs and reflects solar radiation. Such models could be used to make forecasts of the temperature of the Earth's surface and atmosphere. In addition, GISS researchers are working on several more comprehensive atmospheric and oceanic models in an effort to understand and forecast long-term changes in precipitation, air pressure, and wind currents.

Were the scientists able to perfect these ocean atmosphere models, they could get a better handle on how rainfall and cloud cover will change if the Earth warms due to greenhouse gases. Though data collection and modeling efforts have come a long way in the last 25 years, NASA and other agencies dealing with global warming are still a long way off from forecasting the future with certainty. But this new generation of satellite remote sensors is far more sophisticated than their predecessors. Their measurements are more precise and made more frequently over the entire globe every day. Moreover, they can measure a wider range of the Earth's vital signs, enabling scientists to better quantify the key cause-and-effect relationships that drive our climate. Each year dozens of discoveries are made that add to scientists' body of knowledge of how the climate works.

II

One of the key outputs of the 1992 Rio Summit was to highlight and focus attention on the environment, spawning as a result a host of conventions, conferences and other activities related to different environmental issues (collectively called the 'Rio Agreements'). These activities have generated a number of multilateral environment agreements (MEAs), the most recent of them being the Kyoto Protocol promulgated in December 1997. Each of these MEAs require that countries develop specific implementation mechanisms and fulfill obligations involving reporting, training, public education, and other activities. The MEA themes, in fact, lie at the heart of global environmental issues such as CO_2 reduction, eco-efficiency, land degradation, energy systems, technology innovation, etc. Incentive structures—fiscal systems, trade systems and liability systems—have also been proposed as a means of realizing the goals of these MEAs.This working paper focuses on (a) MEAs that are specifically directed at cities, including Local Agenda 21 and Habitat Agenda, and (b) implications of global MEAs on cities. The paper has two components: firstly, it will attempt to build synergies between urban areas and the MEAs in a two way process—how do cities contribute to the conditions and problems addressed by these MEAs? And on the reverse flow, how do these MEAs affect the natural, built-up and social environments of cities? Secondly, it establishs the overlaps, commonalities, inherent relationships and mutual dependencies between these MEAs, focusing on the role of cities and urban stakeholders. It will be grounded in a comprehensive inventory of MEAs with its corresponding urban implications.

Take any of today's environmental problems faced by the inhabitants of this planet, and its causes and pressures can be traced back, directly or indirectly, to cities and its residents' lifestyles and consumption patterns. The forces and processes that constitute 'urban activity' have far-reaching and long-term effects not only on its immediate boundaries, but also on the entire region in which it is positioned. For example, the resources necessary to maintain a city of the size of Greater Tokyo require a land area that is about three and a half time that of Japan as a whole.

Cities and towns in most countries around the world have been gaining considerable attention due to the large number of households migrating to cities and its consequent effects. It has also been due to the centrality of goods and services that cities offer, emerging over the last few decades as the major form of settlement. Proximity to decision-makers and financial markets, large pools of skilled and unskilled workers, and other advantages have made such urban areas the engines of growth for the countries and regions where they are situated. For example, despite the environmental and social problems that it is facing, Bangkok's contribution to the national GDP has been estimated to be more than the combined output of all other cities in Thailand. But population concentration in increasingly smaller land masses has caused a drastic decline in the quality of living both in the residential and work areas. Cities have, in effect, become a barometer of humankind's progress into the 21st century, whether this is an upward or downward trend. Such a scenario has had ripple effects on a variety of sectors such as education, health, labour/job markets, and economic activities.

The concern and problems associated with urban areas and the environment in general have placed such issues high on the agenda of many bilateral and multilateral meetings. The Earth Summit of 1992 in Rio de Janeiro managed to highlight and channel efforts in understanding and acting on environmental problems, making it a key issue to be tackled in trade and commerce, in economic and social development, and in science and technology. Subsequent summits and congresses such as the Social

Summit and the Beijing Conference on Women in 1995, the City Summit/Habitat II in 1996, not to mention innumerable regional, national and local meetings, all have had the larger global environment as an important common denominator in its action plans.

Global Agreements

One of the key outputs of the 1992 Rio Summit was the spawning of a host of conventions, agreements and norms related to different environmental issues (collectively called the 'Rio Agreements'). These activities have generated a number of multilateral environment agreements (MEAs), most recent of them being the Kyoto Protocol on climate change promulgated in Kyoto, Japan December 1997.

Global conventions, agreements and norms have to be concretely adapted and implemented at the local level—which is where real environmental action takes place—targeting the man-on-the-street. It is also important for local issues, priorities, opportunities and experiences to be transmitted back to the global level. Both these cyclical flows, require extensive participation and partnership among all urban stakeholders—local governments, businesses, industry, NGOs, community groups, and ordinary citizens—in cities around the world.

Each MEA require that countries develop specific implementation mechanisms and fulfil obligations involving reporting, training, public education, and other activities . The MEA themes, in fact, lie at the heart of global environmental issues of CO_2 reduction, eco-efficiency, land degradation, eutrofication, energy systems, technology innovation, etc. Incentive structures—fiscal systems, trade systems and liability systems—have also been proposed as a means of realising the goals of these MEAs.

The Urban Areas and Climate Change

The link between COP3 and Urban Environments is a two-way street. Urban areas effect and are affected by climate change (and its primary villain, CO_2) both directly and indirectly. The mega concentration of people and industry in cities are a major contributor to CO_2 emissions. For example, more than 400 vehicles are added to the already congested streets of Bangkok everyday. Urban activities (vehicles, industrial activities, garbage incineration etc.) generate a considerable amount of CO_2. Higher CO_2 levels undoubtedly affect urban dwellers, in terms of respiratory illness, higher temperatures, and a host of related maladies. But 'climate change' is more than just CO_2 levels. It is more than global warming or ozone depletion. Climate change is more than the sum of its parts. Rain patterns may change, climate and agricultural zones may shift towards the poles, melting glaciers and thermal expansion of sea water may raise sea levels, threatening low-lying coastal areas and small islands—the cause-and-effect patterns are endless. Combating the effects of climate change, from an urban perspective, requires a concerted framework within which disparate actions can be positioned and brought together—so that collective effects and efforts can be realised. A cooperative rather than a confrontational approach needs to be adopted, where local actions at the grassroots level add up to a more balanced environmental existence.

The recent Urban Environmental Forum 2000, held in Cape Town in September 2000, raised the following issues in delineating a local dimension to global MEAs:

1. Operational relevance of global agreements, conventions and norms—how well do they fit the operational priorities, needs and capabilities at the local level?

2. Relation to other developmental objectives and priorities—how do the MEAs relate to such issues as poverty alleviation, equity and efficiency?
3. Participation of all relevant stakeholders—what emphasis do the MEAs place on partnerships and alliances, effective and meaningful process of participation, legitimacy of participant stakeholders, trust etc.?
4. The balance of standardization and local differentiation—how are local variations (cultural, economic, geographic) reflected in the norms/ agreements? How are the particular views of city level implementation reconciled with national or global views?
5. Economic consequences of implementation—who gains and who loses as a result of implementing global norms and agreements (the distribution of costs and benefits)?
6. Mixture of tools and mechanisms—how do MEAs facilitate the use of complementary mutually-supporting techniques of implementation; the combination of formal and informal methods; the coherence and consistency of techniques used?
7. Financial and economic incentives—What incentives exist that utilize 'self-interest' and build on market forces to implement MEAs?
8. Defining the roles of different levels of government—how is the authority and responsibility among and between levels of government distributed, relative to city-level implementation?
9. Awareness and understanding—how can (a) the

significance of issues, (b) implications of implementing (and not implementing) MEAs (c) short-term and long-term consequences for different stakeholders be built at the local level?

10. Building political and social commitment—how can commitment be built for MEA implementation at the local level through civil society actors? How can consensus to support implementation be developed and maintained?
11. Institutionalization—How can the MEA consultation processes be formalized at the local level? How institutional structures and/or processes to support local implementation be developed?
12. Sovereignty—how can issues such as ceding sovereign national power, national sovereignty versus city-level cross boundary agreements and actions on MEA implementation, be dealt with effectively?

There is a clear need to focus on MEAs that are specifically directed at cities, including Local Agenda 21 and Habitat Agenda, and delineate implications of other MEAs on cities. Specifically, there is a need to build synergies between urban areas and the MEAs in a two way process—how do cities contribute to the conditions and problems addressed by these MEAs? And on the reverse flow, how do these MEAs affect the natural, built-up and social environments of cities? Secondly, we need to establish the overlaps, commonalties, inherent relationships and mutual dependencies between these MEAs, focusing on the role of cities and urban stakeholders.

The genesis of the idea lies in the fact that effective

implementation of MEAs lies in strong local and urban governments that can take decisive steps at the local level, particularly in urban areas, that have positive, cumulative and global impacts. This calls for a framework of supporting institutions (including the MEA secretariats) at all levels, that support and prop-up urban authorities/local governments. MEA secretariats will have to work in cohort with local governments (or networks or associations of local governments) in generating awareness and disseminating information to link MEA obligations and implementation structures to the policies, programmes and projects that are instituted by local governments. A host of regional, national and local organizations also need to assist the two actors in these processes.

KEY ISSUES THAT AFFECT LOCAL GOVERNMENTS

Quite clearly, there are a number of issues that local governments will need to take into account in catering to global environmental agreements, but also to create a local environment that has positive global impacts. This calls for a clear understanding of the implications of MEAs at the local level; linking action at the local level and understanding their implications at the global level; a model framework of policy and programme development for local governments and authorities; delineation of roles and responsibilities of urban and local stakeholders for MEA implementation and action; and understanding of local capacities and skills that need to be built for the local implementations of global MEAs

- **An understanding of the implications of MEAs at the local level**

As mentioned in the introduction, it is quite clear that the obligations and implication of global and multilateral environmental agreements have definite (but in many cases, not clearly delineated) local dimensions—where both the problems and the solutions start.

- **Linking of action at the local level and their implications at the global level**

The widely accepted mantra of 'Think Global, Act Local' will need to go beyond rhetoric—justifications and clarifications will have to be sought for local environmental action in terms of their direct and indirect global impacts. This linking is critical not only for its impact on the global environment, but also for facilitating participation among local communities, and inspiring action.

- **A model policy framework for local governments in environmental management**

Current environmental management policies at the local level are piecemeal, and scattered, with responsibilities and task allocations made to appropriate departments and agencies among local government entities. An overall framework of policies and programmes, along with their goals/objectives needs to be developed that will allow the placing of actions to be taken, without ignoring goals and objectives of the city as a whole.

- **Delineation of roles and responsibilities of urban and local stakeholders for MEA implementation and action**

In order to operationalize the framework mentioned above, it is very critical to delineate roles and responsibilities of urban and local stakeholders. These will not only help in avoiding overlap and repetition, but will also enable making best use of the strengths and skills of each stakeholder. It will enable the right actors to take the right actions for local environmental management.

- **Understanding of local capacities and skills that need to be built for local implementations of global MEAs**

Each global MEA has a set of unique obligations and actions that need to be taken. But other steps are cross-cutting (for example, a comprehensive 'State of the Urban Environment' report for cities, or a sustainability indicators report), and are necessary for all MEAs. Obligations under each MEA needs to be translated into capacities and skills that needs to be built at the local level.

With more than half the global population now opting to stay in cities and urbanized areas, it is becoming increasingly obvious that cities will play an increasingly important role in shaping the global environment—in terms of both contributing to the problem, but also in actively generating solutions.

Trialogue on Climate Change, Biodiversity and Desertification

Focus on NGOs

outputs of the Rio Summit was to highlight and focus attention on the environment, spawning as a result a host of conventions, conferences and other activities related to myriad environmental issues. The three on-going conventions on Climate Change, Biodiversity, and Desertification have presented a triple challenge both governments and NGOs.

With the objective of adapting a holistic view by integrating the three issues together, The Trialogue is presented as a means of comparing, coordinating and consolidating activities and information related to the three conventions. Focus is placed on providing information on the conventions, and specifically highlighting the activities and actions of NGOs.

Climate Change

UN Framework Convention on Climate Change

Human activities are releasing greenhouse gases into the atmosphere and rising levels of greenhouse gases are expected to cause drastic climate change. The faster the climate changes, the greater will be the risk of damage. Temperatures are expected to raise by 1 to 3.5 deg.C by 2100. The mean sea level is expected to rise 15-95 cm by the year 2100, causing flooding of low-lying areas and other damage. Climatic zones (and thus ecosystems and agricultural zones) could shift towards the poles by 150-550 km in the mid-latitude regions. Forests, deserts, rangelands, and other unmanaged ecosystems would face new climatic stresses. Biodiversity

UN Convention on Biological Diversity

The term "biological diversity" is commonly used to describe the number and variety of living organisms on the planet. It is defined in terms of genes, species, and ecosystems which are the outcome of over 3,000 million years of evolution. The human species depends on biological diversity for its own survival. The Earth's biological resources are vital to humanity's economic and social development. As a result, there is a growing recognition that biological diversity is a global asset of tremendous value to present and future generations. At the same time, the threat to species and ecosystems has never been so great as it is today. Species extinction caused by human activities continues at an alarming rate.

Desertification

UN Convention to Combat Desertification

Land degradation is continuing and increasing at an alarming pace, seriously eroding the world's precious store of productive land. When it happens in the world's drylands it often creates desert-like conditions and is called "desertification". This process happens piecemeal as different areas of degraded land spread and merge together, rather than through advancing desert. The definition agreed by the world's leaders at the 1992 Earth Summit and adopted by the Convention is: "land degradation in arid, semi-arid and sub-humid areas resulting from various factors, including climatic variations and human activities."

Multilateral Environment Agreements (MEAs) and the Urban Arena: Implications for Global Environmental Processes

One of the key outputs of the 1992 Rio Summit was to highlight and focus attention on the environment, spawning as a result a host of conventions, conferences and other activities related to different environmental issues (collectively called the 'Rio Agreements'). These activities have generated a number of multilateral environment agreements (MEAs), the most recent of them being the Kyoto Protocol promulgated in December 1997. Each of these MEAs require that countries develop specific implementation mechanisms and fulfill obligations involving reporting, training, public education, and other activities. The MEA themes, in fact, lie at the heart of global environmental issues such as CO_2 reduction, eco-efficiency, land degradation, energy systems, technology innovation, etc. Incentive structures—fiscal systems, trade systems and liability systems—have also been proposed as a means of realizing the goals of these MEAs.

Many MEAs are specifically directed at cities, including Local Agenda 21 and Habitat Agenda, and other MEAs have clear implications on the way cities function and are managed. There are two sets of questions: how do cities contribute to the conditions and problems addressed by these MEAs? And on the reverse flow, how do these MEAs affect the natural, built-up and social environments of cities? In relation to the above two questions, what are the overlaps, commonalities, inherent relationships and mutual dependencies between these MEAs, within the perspectives of cities and urban stakeholders?

These questions are central to the UEMRI project on "Multilateral Environment Agreements (MEAs) and the Urban Arena."

- Introduction—Multilateral Environment Agreements (MEAs) and the Urban Arena
- Project Plan Summary
- Project Proposal
- Paper: Diversity and Homogeneity: Fostering Innovativeness in Local Environmental Management
- Paper: Tools and Legislation for Environmental Management available to Local Governments
- Paper: Challenges in Strengthening Global-Local Linkages
- Case Study: ICLEI's Cities for Climate Protection
- Case Study: Cities Against Desertification
- Case Study: The Ecological Footprints of Tokyo
- Notes: Participation of Civil Society in the Implementation of MEAs

Additional UEMRI Research Output:

- Trialogue on Climate Change, Biodiversity and Desertification
- Cities and ISO 14001
- Local Agenda 21

Resources on the Internet:

- Resources on Multilateral Environment Agreements (MEAs)
- Selected Global Environmental Assessments
- UNU: Interlinkages Initiative
- UN-HABITAT: Urban Environment Forum (UEF 2000)
- UNEP: International Environmental Governance—Multilateral Environmental Agreements (2001)

6

GLOBAL WARMING AND RENEWABLE ENERGY

Renewable Energy is energy derived from resources that are regenerative or for all practical purposes cannot be depleted. For this reason, renewable energy sources are fundamentally different from fossil fuels, and do not produce as many greenhouse gases and other pollutants as fossil fuel combustion. Mankind's traditional uses of wind, water, and solar energy are widespread in developed and developing countries; but the mass production of electricity using renewable energy sources has become more commonplace recently, reflecting the major threats of climate change, exhaustion of fossil fuels, and the environmental, social and political risks of fossil fuels and nuclear power. Consequently, many countries promote renewable energies through tax incentives and subsidies.

Currently, renewable energy accounts for ca. 14% of the world's energy consumption, but the technical potential is large enough to cover many times current and several times projected energy consumption in 2100. Renewable technologies such as geothermal and hydropower are often economically competitive without subsidies. Other technologies such as solar power is substantially more expensive, although future costs may decline to a fraction of current levels.

Main Renewable Energy Technologies

Renewable energy flows involve natural phenomena such as sunlight, wind, tides and geothermal heat. Each of these sources has unique characteristics which influence how and where they are used. The majority of renewable energy technologies are directly or indirectly powered by the Sun. The Earth-Atmosphere system is in equilibrium such that heat radiation into space is equal to incoming solar radiation, the resulting level of energy within the Earth-Atmosphere system can roughly be described as the Earth's "climate." The hydrosphere (water) absorbs a major fraction of the incoming radiation. Most radiation is absorbed at low latitudes around the equator, but this energy is dissipated around the globe in the form of winds and ocean currents. Wave motion may play a role in the process of transferring mechanical energy between the atmosphere and the ocean through wind stress. Solar energy is also responsible for the distribution of precipitation which is tapped by hydroelectric projects, and for the growth of plants used to create biofuels.

Wind Power

Airflows can be used to run wind turbines and some are capable of producing 5 MW of power. Turbines with rated output of 1.5-3 MW have become the most common for commercial use. The power output of a turbine is a function of the cube of the wind speed, so as wind speed increases, power output increases dramatically. Areas where winds are stronger and more constant, such as offshore and high altitude sites, are preferred locations for wind farms.

Wind power is the fastest growing of the renewable energy technologies. Over the past decade, global installed maximum capacity increased from 2,500 MW in 1992 to just over 40,000 MW at the end of 2003, at an annual growth rate of near 30%. Due to the intermittency of wind resources, most deployed turbines in the EU produce electricity an average of 25% of their rated maximum power (a load factor of 25%), but under favourable wind regimes some reach 35% or higher. The load factor is generally higher in winter. It would mean that a typical 5 MW turbine in the EU would have an average output of 1.7 MW.

Globally, the long-term technical potential of wind energy is believed to be five times current production global energy consumption or 40 times current electricity demand. This could require large amounts of land to be utilized for wind turbines, particularly in areas of higher wind resources. Offshore resources experience mean wind speeds of ~90% greater than that of land, so offshore resources could contribute substantially more energy. This number could also increase with higher altitude ground-based or airborne wind turbines.

Wind strengths near the Earth's surface vary and thus cannot guarantee continuous power unless combined with other energy sources or storage systems. Some estimates suggest that 1,000 MW of conventional wind generation capacity can be relied on for just 333 MW of continuous power. While this might change as technology evolves, advocates have suggested incorporating wind power with other power sources, or the use of energy storage techniques, with this in mind. It is best used in the context of a system that has significant reserve capacity such as hydro, or reserve load, such as a desalination plant, to mitigate the economic effects of resource variability.

Wind power is renewable and produces no greenhouse gases during operation, such as carbon dioxide and methane. Studies of birds and offshore wind farms in Europe have found that there are very few bird collisions. Several offshore wind sites in Europe have been in areas heavily used by seabirds. Improvements in wind turbine design, including a much slower rate of rotation of the blades and a smooth tower base instead of perchable lattice towers, have helped reduce bird mortality at wind farms around the world. However older smaller wind turbines may be hazardous to flying birds.[*citation needed*] Birds are severely impacted by fossil fuel energy; examples include birds dying from exposure to oil spills, habitat loss from acid rain and mountaintop removal coal mining, and mercury poisoning.

Water Power

Energy in water (in the form of motive energy or temperature differences) can be harnessed and used. Since water is about a thousand times denser than air, even a slow flowing stream of water, or moderate sea swell, can yield considerable amounts of energy.

There are many forms of water energy:

- Hydroelectric energy is a term usually reserved for large-scale hydroelectric dams.
- Micro hydro systems are hydroelectric power installations that typically produce up to 100 kW of power. They are often used in water rich areas as a Remote Area Power Supply (RAPS). There are many of these installations around the world, including several delivering around 50 kW in the Solomon Islands.

- Wave power uses the energy in waves. The waves will usually make large pontoons go up and down in the water, leaving an area with reduced wave height in the "shadow". Wave power has now reached commercialization.
- Tidal power captures energy from the tides in a vertical direction. Tides come in, raise water levels in a basin, and tides roll out. Around low tide, the water in the basin is discharged through a turbine.
- Tidal stream power captures energy from the flow of tides, usually using underwater plant resembling a small wind turbine. Tidal stream power demonstration projects exist, but large scale development requires additional capital.
- Ocean thermal energy conversion (OTEC) uses the temperature difference between the warmer surface of the ocean and the colder lower recesses. To this end, it employs a cyclic heat engine. OTEC has not been field-tested on a large scale.
- Deep lake water cooling, although not technically an energy generation method, can save a lot of energy in summer. It uses submerged pipes as a heat sink for climate control systems. Lake-bottom water is a year-round local constant of about 4 °C.
- Blue energy is the reverse of desalination. This form of energy is in research.

Solar Energy Use

In this context, "solar energy" refers to energy that is collected from sunlight. Solar energy can be applied in many ways, including to:

- Generate electricity using photovoltaic solar cells.
- Generate electricity using concentrated solar power.
- Generate electricity by heating trapped air which rotates turbines in a Solar updraft tower.
- Heat buildings, directly, through passive solar design.
- Heat foodstuffs, through solar ovens.
- Heat water or air for domestic hot water and space heating needs using solar-thermal panels.
- Heat and cool air through use of solar chimneys.

Biofuel: Plants use photosynthesis to grow and produce biomass. Also known as biomatter, biomass can be used directly as fuel or to produce liquid biofuel. Agriculturally produced biomass fuels, such as biodiesel, ethanol and bagasse (often a by-product of sugar cane cultivation) can be burned in internal combustion engines or boilers. Typically biofuel is burned to release its stored chemical energy. Research into more efficient methods of converting biofuels and other fuels into electricity utilizing fuel cells is an area of very active work.

Liquid Biofuel

Liquid biofuel is usually either a bioalcohol such as ethanol or a bio-oil such as biodiesel and straight vegetable oil. Biodiesel can be used in modern diesel vehicles with little or no modification to the engine and can be made from waste and virgin vegetable and animal oil and fats (lipids). Virgin vegetable oils can be used in modified diesel engines. In fact the Diesel engine was originally designed to run on vegetable oil rather than fossil fuel. A major benefit of biodiesel is lower emissions. The use of biodiesel reduces emission of carbon monoxide and other hydrocarbons by 20 to 40%. In some areas corn, cornstalks, sugarbeets, sugar cane, and switchgrasses are grown specifically to produce ethanol (also known as grain alcohol) a liquid which can be used in internal combustion engines and fuel cells. Ethanol is being phased into the current energy infrastructure. E85 is a fuel composed of 85% ethanol and 15% gasoline that is sold to consumers.

In the future, there might be bio-synthetic liquid fuel available. It can be produced by the Fischer-Tropsch process, also called Biomass-To-Liquids (BTL).

Solid Biomass: Direct use is usually in the form of combustible solids, either wood, the biogenic portion of municipal solid waste or combustible field crops. Field crops may be grown specifically for combustion or may be used for other purposes, and the processed plant waste then used for combustion. Most sorts of biomatter, including dried manure, can actually be burnt to heat water and to drive turbines.

Sugar cane residue, wheat chaff, corn cobs and other plant matter can be, and is, burnt quite successfully. The net Carbon Dioxide emissions that are added to the atmosphere by this process are only from the fossil fuel that is consumed to plant, fertilize, harvest and transport the biomass. Processes to grow perenials such as switchgrass, miscanthus, and willow, field pelletize and co-fire with coal for electricity generation are being studied and appear to be economically viable. Co-firing this cellulosic biomass with coal to make electricity is more effective for reducing carbon dioxide emmissions to the atmosphere than using it to make ethanol.

Solid biomass can also be gasified, and used as described in the next section.

Biogas: Biogas can easily be produced from current waste streams, such as: paper production, sugar production, sewage, animal waste and so forth. These various waste streams have to be slurried together and allowed to naturally ferment, producing methane gas. This can be done by converting current sewage plants into biogas plants. When a biogas plant has extracted all the methane it can, the remains are sometimes better suitable as fertilizer than the original biomass.

Alternatively biogas can be produced via advanced waste processing systems such as mechanical biological treatment. These systems recover the recyclable elements of household waste and process the biodegradable fraction in anaerobic digesters.

Renewable natural gas is a biogas which has been upgraded to a quality similar to natural gas. By upgrading the quality to that of natural gas, it becomes possible to distribute the gas to the mass market via gas grid.

Geothermal Energy

Geothermal energy is energy obtained by tapping the heat of the earth itself, usually from kilometers deep into the Earth's crust. It is expensive to build a power station but operating costs are low resulting in low energy costs for suitable sites. Ultimately, this energy derives from the radioactive decay in the core of the Earth, which heats the Earth from the inside out.

Three types of power plants are used to generate power from geothermal energy: dry steam, flash, and binary. Dry steam plants take steam out of fractures in the ground and use it to directly drive a turbine that spins a generator. Flash plants take hot water, usually at temperatures over 200 °C, out of the ground, and allows it to boil as it rises to the surface then separates the steam phase in steam/water separators and then runs the steam through a turbine. In binary plants, the hot water flows through heat exchangers, boiling an organic fluid that spins the turbine. The condensed steam and remaining geothermal fluid from all three types of plants are injected back into the hot rock to pick up more heat.

Although geothermal sites are capable of providing heat for many decades, eventually they are depleted as the ground cools. The government of Iceland states *It should be stressed that the geothermal resource is not strictly renewable in the same sense as the hydro resource.* It estimates that Iceland's geothermal energy could provide 1700 MW for over 100 years, compared to the current production of 140 MW.

The geothermal energy from the core of the Earth is closer to the surface in some areas than in others. Where hot underground steam or water can be tapped and brought to the surface it may be used to generate electricity. Such geothermal power sources exist in certain geologically unstable parts of the world such as Iceland, New Zealand, United States, the Philippines and Italy. The two most prominent areas for this in the United States are in the Yellowstone basin and in northern California. Iceland produced 170 MW geothermal power and heated 86% of all houses in the year 2000 through geothermal energy. Some 8000 MW of capacity is operational in total.

The is also the potential to generate geothermal energy from Hot Dry Rocks. Holes at least 3km deep are drilled into the earth. Some of these holes pump water into the earth, while other holes pump hot water out. The heat resource consists of hot underground radiogenic granite rocks, which heat up when there is enough sediment between the rock and the earths surface. Several companies in Australia are exploring this technology.

Potential

While currently renewable energy sources only supply a modest fraction of current energy use (ca. 14% of primary energy use, mostly from traditional biomass), there is a much potential that could be exploited in the future. As the table below illustrates, the technical potential of renewable energy sources is more than 18 times current global primary energy use and furthermore several times higher than projected energy use in 2100.

The Renewable Energy Resource Base (Exajoules a year)

	Current use (2001)	*Technical potential*	*Theoretical potential*
Hydropower	9	50	147
Biomass energy	50	>276	2,900
Solar energy	0.1	>1,575	3,900,000
Wind energy	0.12	640	6,000
Geothermal energy	0.6	5,000	140,000,000
Ocean energy	not estimated	not estimated	7,400
Total	60	>7,600	>144,000,000

Current use is in primary energy equivalent.

For comparison, the current global primary energy use (2001) is 402 Exajoules a year.

Source: World Energy Assessment 2001.

There are many different ways to assess potentials. The theoretical potential indicates the amount of energy theoretically available for energy purposes, such as, in the case of solar power, the amount of incoming radiation at the earth's surface. The technical potential is a more practical estimate of how much could be put to human use by considering conversion efficiencies of the available technology and available land area. To give an idea of the constraints, the estimate for solar energy assumes that 1% of the world's unused land surface is used for solar power.

The technical potentials generally do not include economic or other environmental constraints, and the potentials that could be realized at an economically competitive level under current conditions and in a short time-frame is lower still. It should also be noted that intermittent sources such as wind, solar, tidal, and wave energy may eventually require energy storage and/or back-up to guarantee reliable supply at large penetrations.

Renewable Energy Commercialization

Costs

Renewable energy technologies encompass a broad, diverse array of technologies, and the current status of these different technologies varies considerably. Some technologies are already mature and economically competitive (e.g. geothermal and hydropower), other technologies need additional development steps to become competitive without subsidies.

The table below shows an overview of costs of various renewable energy technologies. For comparison with the prices in the table, electricity production from a conventional coal-fired plant costs about 4c/kWh. Achieving further cost reductions as indicated in the table below requires further technology development, market deployment, and an increase in production capacities to mass production levels.

	2001 energy costs	*Potential future energy cost*
Electricity		
Wind	4-8 ¢/kWh	3-10 ¢/kWh
Solar photovoltaic	25-160 ¢/kWh	5-25 ¢/kWh
Solar thermal	12-34 ¢/kWh	4-20 ¢/kWh
Large hydropower	2-10 ¢/kWh	2-10 ¢/kWh
Small hydropower	2-12 ¢/kWh	2-10 ¢/kWh
Geothermal	2-10 ¢/kWh	1-8 ¢/kWh
Biomass	3-12 ¢/kWh	4-10 ¢/kWh
Heat		
Geothermal heat	0.5-5 ¢/kWh	0.5-5 ¢/kWh
Biomass—heat	1-6 ¢/kWh	1-5 ¢/kWh
Low temperature solar heat	2-25 ¢/kWh	2-10 ¢/kWh

All costs are in 2001 $-cent per kilowatt-hour.Source: World Energy Assessment, 2004 update.

Wind Power Market Grows

The Global Wind Energy Council (GWEC) show that 2006 recorded an increase of installed capacity for 15,197 megawatts (MW), taking the total installed wind energy capacity to 74,223 MW, up from 59,091 MW in 2005.

Despite constraints facing supply chains for wind turbines, the annual market for wind continued to increase at the rate of 32% following the 2005 record year, in which the market grew by 41%.

In terms of economic value, the wind energy sector has become one of the important players in the energy markets, with the total value of new generating equipment installed in 2006 reaching Euro 18 billion, or US$23 billion.

The countries with the highest total installed capacity are Germany (20,621 MW), Spain (11,615 MW), the USA (11,603 MW), India (6,270 MW) and Denmark (3,136). In terms of new installed capacity in 2006, the US lead with 2,454 MW, followed by Germany (2,233 MW), India (1,840 MW), Spain (1,587 MW), China (1,347 MW) and France (810 MW).

In the UK, a licence to build the world's largest offshore windfarm, in the Thames estuary, has been granted. The London Array windfarm, 12 miles off Kent and Essex, should eventually consist of 341 turbines, occupying an area of 90 square miles. This is a £1.5 billion, 1,000 megawatt project, which will power one-third of London homes. The windfarm will produce an amount of energy that, if generated by conventional means, would result in 1.9 million tonnes of carbon dioxide emissions every year. It could also make up to 10% of the Government's 2010 renewables target.

New Generation of Solar Thermal Plants

Construction of the largest solar thermal power plant to be built in 15 years, in Boulder City, Nevada, is nearly complete.

The 64MW Nevada Solar One power plant will generate enough power to meet the electricity needs of about 40,000 households and follows in the steps of the 354MW SEGS solar thermal power plants located in California's Mojave Desert. While California's solar plants have generated billions of kilowatt hours of electricity for the past two decades, the Nevada Solar One plant will use new technologies to capture even more energy from the sun.

The California Solar Initiative

As part of Governor Arnold Schwarzenegger's Million Solar Roofs Program, California has set a goal to create 3,000 megawatts of new, solar-produced electricity by 2017—moving the state toward a cleaner energy future and helping lower the cost of solar systems for consumers. This is a comprehensive $2.8 billion program.

The California Solar Initiative offers cash incentives on solar PV systems of up to $2.50 a watt. These incentives, combined with federal tax incentives, can cover up to 50% of the total cost of a solar panel system. It should also be noted that there are many financial incentives to support the use of renewable energy in other US states.

Construction of a 40 MW solar generation power plant is underway in the Saxon region of Germany. The Waldpolenz Solar Park will consist of some 550,000 thin-film solar modules. The direct current produced in the modules will be converted into alternating current and fed completely into the power grid. Once completed in 2009, the project will be one of the largest photovoltaic projects ever constructed. Currently the biggest PV plant in the world has an output capacity of around 12 megawatts.

A large photovoltaic power project has been completed in Portugal, in one of the Europe's sunniest areas. The 11 megawatt plant covers 150 acres and is comprised of 52,000 PV panels. The panels are raised 2 metres off the ground and the area will remain productive grazing land. The project will provide enough energy for 8,000 homes and will save an estimated 30,000 tonnes of carbon dioxide emissions per year.

A $420 million large-scale Solar power station in Victoria is to be the biggest and most efficient solar photovoltaic power station in the world. Australian company Solar Systems will demonstrate its unique, world leading design incorporating space technology in a 154MW solar power station connected to the national grid. The power station will have the capability to concentrate the sun by 500 times onto the solar cells for ultra high power output. The Victorian power station will generate clean electricity directly from the sun to meet the annual needs of over 45,000 homes with zero greenhouse gas emissions.

However, when it comes to renewable energy systems and PV, it is not just large systems that matter. Building-integrated photovoltaics or "onsite" PV systems have the advantage of being matched to end use energy needs in terms of scale. So the energy is supplied close to where it is needed.

Onsite Renewable Technologies

To promote energy efficiency and environmentally sensitive energy generation, EPA facilities in the United States are using renewable energy technologies to supplement or replace a large portion of their energy requirements at the following facilities:

- Ada, Oklahoma (geothermal heat pump)
- Ann Arbor, Michigan (fuel cell)
- Chicago, Illinois, Regional Office (photovoltaic array)
- Corvallis, Oregon (photovoltaic array)
- Edison, New Jersey (solar water heating)
- Gulf Breeze, Florida (solar lighting)
- Golden, Colorado (wind power and transpired solar collector)

- Manchester, Washington (wind power)
- Research Triangle Park, North Carolina (photovoltaic solar panels and street lights)

Use of Ethanol Blends for Transportation in the U.S.

Most cars on the road today in the U.S. can run on blends of up to 10% ethanol, and motor vehicle manufacturers already produce vehicles designed to run on much higher ethanol blends. Ford, DaimlerChrysler, and GM are among the automobile companies that sell "flexible-fuel" cars, trucks, and minivans that can use gasoline and ethanol blends ranging from pure gasoline up to 85% ethanol (E85). By mid-2006, there were approximately six million E85-compatible vehicles on U.S. roads.

The challenge is to expand the market for biofuels beyond the farm states where they have been most popular to date. Flex-fuel vehicles are assisting in this transition because they allow drivers to choose different fuels based on price and availability. The *Energy Policy Act of 2005*, which calls for 7.5 billion gallons of biofuels to be used annually by 2012, will also help to expand the market.

Portugal now has the world's first commercial wave farm, the *Aguçadora Wave Park*, established in 2006. The farm will initially use three Pelmis P-750 machines generating 2.25 MW. Initial costs are put at 8,5 million euro. Subject to successful operation, a further 70 million euro is likely to be invested before 2009 on a further 28 machines to generate 525 MW. Funding for a wave farm in Scotland was announced in February, 2007 by the Scottish Executive, at a cost of over 4 million pounds, as part of a £13 million funding packages for ocean power in Scotland. The farm will be the world's largest with a capacity of 3MW generated by four Pelamis machines.

Geothermal Energy Prospects

By the end of 2005 worldwide use of geothermal energy for electricity had reached 9.3 GWs, with an additional 28 GW used directly for heating. If heat recovered by ground source heat pumps is included, the non-electric use of geothermal energy is estimated at more than 100 GWt (gigawatts of thermal power) and is used commercially in over 70 countries. During 2005 contracts were placed for an additional 0.5 GW of capacity in the United States, while there were also plants under construction in 11 other countries.

Conclusion

Critics suggest that some renewable energy applications may create pollution, be dangerous, take up large amounts of land, or be incapable of generating a large net amount of energy. Proponents advocate the use of "appropriate renewables", also known as soft energy technologies, as these have many advantages.

Availability

There is no shortage of solar-derived energy on Earth. Indeed the storages and flows of energy on the planet are very large relative to human needs.

- The amount of solar energy intercepted by the Earth every minute is greater than the amount of energy the world uses in fossil fuels each year.
- Tropical oceans absorb 560 trillion gigajoules (GJ) of solar energy each year, equivalent to 1,600 times the world's annual energy use.
- The energy in the winds that blow across the United States each year could produce more than 16 billion GJ of electricity—more than one

and one-half times the electricity consumed in the United States in 2000.

- Annual photosynthesis by the vegetation in the United States is 50 billion GJ, equivalent to nearly 60% of the nation's annual fossil fuel use.

Yet a recurring criticism of some renewable sources is their intermittent nature. But a variety of renewable sources in combination can overcome this problem. As Amory Lovins explains:

> "Stormy weather, bad for direct solar collection, is generally good for windmills and small hydropower plants; dry, sunny weather, bad for hydropower, is ideal for photovoltaics.

The challenge of variable power supply may be further alleviated by energy storage. Available storage options include pumped-storage hydro systems, batteries, hydrogen fuel cells, and thermal mass. Initial investments in such energy storage systems can be high, although the costs can be recovered over the life of the system.

Wave energy is continuously available, although wave intensity varies by season. A wave energy scheme installed in Australia generates electricity with an 80% availability factor.

Reliability

Renewable energy sources are often dismissed as unreliable. Yet a diversity of renewable sources, each serving fewer and nearer users, would also greatly restrict the area blacked out if a grid connecting them failed. And when renewable energy sources do fail, they generally fail for shorter periods than do large power plants.

Our complex, interdependent systems for the production and delivery of energy are vulnerable to simple but devastating acts of sabotage and terrorism. More efficient, diverse, dispersed, renewable energy systems can make major failures impossible.

Storage of energy from renewable energy systems can also contribute to improved reliability.

Aesthetics

Some people dislike the aesthetics of large solar-electric installations outside cities. However, methods and opportunities exist to deploy these renewable technologies in an efficient and aesthetically pleasing way: fixed solar collectors can double as noise barriers along highways; tremendous roadway, parking lot, and roof-top area is available already (and rooftops could even be replaced totally by solar collectors); amorphous photovoltaic cells can be used to tint windows and produce energy, etc.

Environmental and Social Considerations

While most renewable energy sources do not produce pollution directly, the materials, industrial processes, and construction equipment used to create them may generate waste and pollution. Some renewable energy systems actually create environmental problems. For instance, older wind turbines can be hazardous to flying birds.

Land Area Required

Another environmental issue, particularly with biomass and biofuels, is the large amount of land required to harvest energy, which otherwise could be used for other purposes or left as undeveloped land. However, it should be pointed out that these fuels may reduce the need for harvesting non-renewable energy sources, such as vast strip-mined areas and slag mountains for coal, safety zones around nuclear plants, and hundreds of square miles being strip-mined for oil sands.

In the U.S., crops grown for biofuels are the most land- and water-intensive of the renewable energy sources. In 2005, about 12% of the nation's corn crop (covering 11 million acres (45,000 km^2) of farmland) was used to produce four billion gallons of ethanol—which equates to about 2% of annual U.S. gasoline consumption. For biofuels to make a much larger contribution to the energy economy, the industry will have to accelerate the development of new feedstocks, agricultural practices, and technologies that are more land and water efficient. Already, the efficiency of biofuels production has increased significantly and there are new methods to boost biofuel production.

More generally, renewable energy as a whole is sometimes viewed as too land-intensive to be practical. Yet harnessing renewable energy for electricity production requires less land and water than does our current energy system. Solar power plants that concentrate sunlight in desert areas require 10 km^2 per GWh (2,540 acre/GWh). On a lifecycle basis, this is less land than a comparable coal or hydropower plant requires, and because most deserts are sparsely populated, there is plenty of room for solar power plants. A little over 4,000 square miles—equivalent to 3.4% of the land in New Mexico—would be sufficient to produce 30% of the United States' electricity.

Hydroelectric Dams

The major advantage of hydroelectric systems is the elimination of the cost of fuel. Other advantages include longer life than fuel-fired generation, low operating costs, and the provision of facilities for water sports. Operation of pumped-storage plants improves the daily load factor of the generation system. Overall, hydroelectric power can be far less expensive than electricity generated from fossil fuels or nuclear energy, and areas with abundant hydroelectric power attract industry.

However, there are several major disadvantages of hydroelectric systems. These include: dislocation of people living where the reservoirs are planned, release of significant amounts of carbon dioxide at construction and flooding of the reservoir, disruption of aquatic ecosystems and birdlife, adverse impacts on the river environment, potential risks of sabotage and terrorism, and in rare cases catastrophic failure of the dam wall.

Hydroelectric power is now more difficult to site in developed nations because most major sites within these nations are either already being exploited or may be unavailable for other reasons such as environmental considerations.

Wind Farms

A wind farm, when installed on agricultural land, has one of the lowest environmental impacts of all energy sources:

- It occupies less land area per kilowatt-hour (kWh) of electricity generated than any other energy conversion system, apart from rooftop solar energy, and is compatible with grazing and crops.
- It generates the energy used in its construction in just 3 months of operation, yet its operational lifetime is 20-25 years.
- Greenhouse gas emissions and air pollution produced by its construction are tiny and declining.

There are no emissions or pollution produced by its operation.

- In substituting for base-load coal power, wind power produces a net decrease in greenhouse gas emissions and air pollution, and a net increase in biodiversity.
- Modern wind turbines are almost silent and rotate so slowly (in terms of revolutions per minute) that they are rarely a hazard to birds.

Longevity Issues

Though a source of renewable energy may last for billions of years, renewable energy infrastructure, like hydroelectric dams, will not last forever, and must be removed and replaced at some point. Events like the shifting of riverbeds, or changing weather patterns could potentially alter or even halt the function of hydroelectric dams, lowering the amount of time they are available to generate electricity.

Although geothermal sites are capable of providing heat for many decades, eventually specific locations may cool down. It is likely that in these locations, the system was designed too large for the site, since there is only so much energy that can be stored and replenished in a given volume of earth. Some interpret this as meaning a specific geothermal location can undergo depletion, and question whether geothermal energy is truly renewable. The government of Iceland states "it should be stressed that the geothermal resource is not strictly renewable in the same sense as the hydro resource." It estimates that Iceland's geothermal energy could provide 1700 MW for over 100 years, compared to the current production of 140 MW. The International Energy Agency classifies geothermal power as renewable.

Biofuels Net Energy Balance

All biomass needs to go through some of these steps: it needs to be grown, collected, dried, fermented and burned. All of these steps require resources and an infrastructure.

Opponents of corn ethanol production in the U.S. often quote the 2005 paper of David Pimentel, a retired Entomologist, and Tadeusz Patzek, a Geological Engineer from Berkeley. Both have been exceptionally critical of ethanol and other biofuels. Their studies contend that ethanol, and biofuels in general, are "energy negative", meaning they take more energy to produce than is contained in the final product. A 2006 report by the U.S. Department Agriculture compared the methodologies used by a number of researchers on this subject and found that the majority of researchers think the energy balance for ethanol is positive. In fact, a large number of recent studies, including a 2006 article in the prestigious journal *Science* offer the consensus opinion that fuels like ethanol are energy positive. Furthermore, it should be pointed out that fossil fuels also require significant energy inputs which have seldom been accounted for in the past.

It should also be noted that the growing ethanol and biodiesel industries are providing jobs in plant construction, operations, and maintenance, mostly in rural communities. According to the Renewable Fuels Association, the ethanol industry created almost 154,000 U.S. jobs in 2005 alone, boosting household income by $5.7 billion. It also contributed about $3.5 billion in tax revenues at the local, state, and federal levels.

According to the International Energy Agency, new biofuels technologies being developed today, notably lignocellulosic ethanol, could allow biofuels to play a much bigger role in the future than previously thought.

Concluding Comment

The U.S. electric power industry now relies on large, central power stations, including coal, natural gas, nuclear, and hydropower plants that together generate more than 95% of the nation's electricity. Over the next few decades uses of renewable energy could help to diversify the nation's bulk power supply. Already, appropriate renewable resources (which excludes large hydropower) produce 12% of northern California's electricity.

Although most of today's electricity comes from large, central-station power plants, new technologies offer a range of options for generating electricity nearer to where it is needed, saving on the cost of transmitting and distributing power and improving the overall efficiency and reliability of the system.

Improving energy efficiency represents the most immediate and often the most cost-effectiveway to reduce oil dependence, improve energy security, and reduce the health and environmental impact of the energy system. By reducing the total energy requirements of the economy, improved energy efficiency will make increased reliance on renewable energy sources more practical and affordable.

Other Issues

Fossil Fuels

Z though not universally held, Western (biogenic) theory is that fossil fuels are the altered remnants of ancient plant and animal life, deposited in sedimentary rocks millions of years ago, which have rested underground, mostly dormant, since that time. Although this process may continue today, it is extremely slow and produces a negligible amount of these resources compared to the rate of consumption by humans. Therefore, the Earth will eventually run out of fossil fuels (*see peak oil*). Fossil fuels are therefore not considered a renewable energy source, and are often contrasted with renewables in the context of future energy development.

Many believe that a move away from a US economy that is primarily dependent on fossil fuels will allow a more even-handed approach to foreign policy. Former CIA Director James Woolsey recently outlined the national security arguments in favor of moving away from fossil fuels.

The coal industry in the US is publicly claiming that coal is a renewable energy source because coal was originally biomass.Some scientists hold the view that the formation of fossil fuels was a one-time event, made possible by unique conditions during the Devonian period, such as increased oxygen levels and huge swamps.

Transmission

If renewable and distributed generation were to become widespread, electric power transmission and electricity distribution systems might no longer be the main distributors of electrical energy but would operate to balance the electricity needs of local communities. Those with surplus energy would sell to areas needing "top ups". That is, network operation would require a shift from 'passive management' — where generators are hooked up and the system is operated to get electricity 'downstream' to the consumer — to 'active management', wherein generators are spread across a network and inputs and outputs need to be constantly monitored to ensure proper balancing occurs within the system. Some Governments and regulators are moving to address this, though much remains to be done. One potential solution is the increased use of active management of electricity transmission and distribution networks. This will require significant changes in the way that such networks are operated.

However, on a smaller scale, use of renewable energy that can often be produced "onsite" lowers the requirements electricity distribution systems have to fulfill. Current systems, while rarely economically efficient, have proven an average household with a solar panel array and energy storage system of the right size needs electricity from outside sources for only a few hours every week. By matching electricity supply to end-use needs in this way, advocates of renewable energy and the soft energy path believe electricity systems will become smaller and easier to manage, rather than the opposite.

Renewable heat is an application of renewable energy, namely the generation of heat from renewable sources. In some cases, contemporary discussion on renewable energy focuses on the generation of electrical, rather than heat energy. This is despite the fact that many colder countries consume more energy for heating than as electricity. On an annual basis the United Kingdom consumes 350 TWh of electric power, and 840 TWh of gas and other fuels for heating. The residential sector alone consumes a massive 550 TWh of energy for heating, mainly in the form of gas.

Renewable electric power is becoming cheap and convenient enough to place it, in many cases, within reach of the average consumer. By contrast, the market for renewable heat is mostly inaccessible to domestic consumers due to inconvenience of supply, and high capital costs.

Heating accounts for a large proportion of energy consumption, however a universally accessible market for renewable heat is yet to emerge. Solutions such as geothermal heat pumps may be more widely applicable, but may not be economical in all cases. Also see renewable energy development.

Kerosene, a petroleum-based fuel currently sourced from non-renewable sources, is considered to be the only fuel practical and economic for commercial jet-engine aviation. However, kerosene can now be manufactured from the light crude oil that is the output of the Thermal depolymerization of renewable feedstocks. Biodiesel, another candidate aviation fuel, is problematic due its tendency to freeze more readily than kerosene. Smaller piston-engined aircraft are mainly fueled by aviation grade gasoline (avgas) but are increasingly being fueled by ethanol or diesel. Given the proper equipment to prevent fuel gelling, a diesel-powered piston aircraft engine can be powered efficiently by biodiesel.

REFERENCES

EWEA Executive summary Analysis of Wind Energy in the EU-25. European Wind Energy Association. Retrieved on 2007-03-11.

Potentials and costs for renewable electricity generation

Renewable Energy, Sorensen, Elsevier 2004

Response of Wairakei Geothermal Reservoir to 40 Years of Production, 2006 (pdf) Allan Clotworthy, Proceedings World Geothermal Congress 2000. (accessed 30 March)

Status And Perspectives of Biomass-To-Liquid Fuels in the European Union

United States Department of Energy Glossary of energy-related terms (URL accessed Dec 21, 2006)

World Energy Assessment 2001, Chapter 5: Energy Resources (table 5.26), coordinating lead author Hans-Holger Rogner. Available for download at its UNDP site.

World Energy Assessment 2004 Update, (energy costs from Table 7). Available for download at its UNDP site.

World Energy Outlook 2004; OECD/International Energy Agency 2004.

APPENDICES

[Since climate-change has comeup in big way affecting the live (both fauna & flora on east adversely. Kyoto Protocol is a milestone in checking man-made causes of global warming. Her we have been reproducing the original text for the perusal of academics, research policy planners and activists in the field. —Ed.]

APPENDIX I

KYOTO PROTOCOL TO THE UNITED NATIONS FRAMEWORK CONVENTION ON CLIMATE CHANGE

The Parties to this Protocol,

Being *Parties to the United Nations Framework Convention on Climate Change, hereinafter referred to as "the Convention",*

In pursuit *of the ultimate objective of the Convention as stated in its Article 2,*

Recalling *the provisions of the Convention,*

Being guided *by Article 3 of the Convention,*

Pursuant to *the Berlin Mandate adopted by decision 1/ CP.1 of the*

Conference of the Parties to the Convention at its first session,

Have agreed as follows:

Article 1

For the purposes of this Protocol, the definitions contained in Article 1 of the Convention shall apply. In addition:

1. "Conference of the Parties" means the Conference of the Parties to the Convention.

2. "Convention" means the United Nations Framework Convention on Climate Change, adopted in New York on 9 May 1992.

3. "Intergovernmental Panel on Climate Change" means the Intergovernmental Panel on Climate Change established in 1988 jointly by the World Meteorological Organization and the United Nations Environment Programme.

4. "Montreal Protocol" means the Montreal Protocol on Substances that Deplete the Ozone Layer, adopted in Montreal on 16 September 1987 and as subsequently adjusted and amended.

5. "Parties present and voting" means Parties present and casting an affirmative or negative vote.

6. "Party" means, unless the context otherwise indicates, a Party to this Protocol.

7. "Party included in Annex I" means a Party included in Annex I to the Convention, as may be amended, or a Party which has made a notification under Article 4, paragraph 2(g), of the Convention.

Article 2

1. Each Party included in Annex I, in achieving its quantified emission limitation and reduction commitments

under Article 3, in order to promote sustainable development, shall:

(a) Implement and/or further elaborate policies and measures in accordance with its national circumstances, such as:

   (i) Enhancement of energy efficiency in relevant sectors of the national economy;

   (ii) Protection and enhancement of sinks and reservoirs of greenhouse gases not controlled by the Montreal Protocol, taking into account its commitments under relevant international environmental agreements; promotion of sustainable forest management practices, afforestation and reforestation;

   (iii) Promotion of sustainable forms of agriculture in light of climate change considerations;

   (iv) Research on, and promotion, development and increased use of, new and renewable forms of energy, of carbon dioxide sequestration technologies and of advanced and innovative environmentally sound technologies;

   (v) Progressive reduction or phasing out of market imperfections, fiscal incentives, tax and duty exemptions and subsidies in all greenhouse gas emitting sectors that run counter to the objective of the Convention and application of market instruments;

   (vi) Encouragement of appropriate reforms in relevant sectors aimed at promoting policies and measures which limit or reduce emissions of greenhouse gases not controlled by the Montreal Protocol;

(vii) Measures to limit and/or reduce emissions of greenhouse gases not controlled by the Montreal Protocol in the transport sector;

(viii) Limitation and/or reduction of methane emissions through recovery and use in waste management, as well as in the production, transport and distribution of energy;

(b) Cooperate with other such Parties to enhance the individual and combined effectiveness of their policies and measures adopted under this Article, pursuant to Article 4, paragraph 2(e)(i), of the Convention. To this end, these Parties shall take steps to share their experience and exchange information on such policies and measures, including developing ways of improving their comparability, transparency and effectiveness. The Conference of the Parties serving as the meeting of the Parties to this Protocol shall, at its first session or as soon as practicable thereafter, consider ways to facilitate such cooperation, taking into account all relevant information.

2. The Parties included in Annex I shall pursue limitation or reduction of emissions of greenhouse gases not controlled by the Montreal Protocol from aviation and marine bunker fuels, working through the International Civil Aviation Organization and the International Maritime Organization, respectively.

3. The Parties included in Annex I shall strive to

implement policies and measures under this Article in such a way as to minimize adverse effects, including the adverse effects of climate change, effects on international trade, and social, environmental and economic impacts on other Parties, especially developing country Parties and in particular those identified in Article 4, paragraphs 8 and 9, of the Convention, taking into account Article 3 of the Convention. The Conference of the Parties serving as the meeting of the Parties to this Protocol may take further action, as appropriate, to promote the implementation of the provisions of this paragraph.

4. The Conference of the Parties serving as the meeting of the Parties to this Protocol, if it decides that it would be beneficial to coordinate any of the policies and measures in paragraph 1(a) above, taking into account different national circumstances and potential effects, shall consider ways and means to elaborate the coordination of such policies and measures.

Article 3

1. The Parties included in Annex I shall, individually or jointly, ensure that their aggregate anthropogenic carbon dioxide equivalent emissions of the greenhouse gases listed in Annex A do not exceed their assigned amounts, calculated pursuant to their quantified emission limitation and reduction commitments inscribed in Annex B and in accordance with the provisions of this Article, with a view to reducing their overall emissions of such gases by at least 5 per cent below 1990 levels in the commitment period 2008 to 2012.

2. Each Party included in Annex I shall, by 2005, have made demonstrable progress in achieving its commitments under this Protocol.

3. The net changes in greenhouse gas emissions by sources and removals by sinks resulting from direct human-induced land-use change and forestry activities, limited to afforestation, reforestation and deforestation since 1990, measured as verifiable changes in carbon stocks in each commitment period, shall be used to meet the commitments under this Article of each Party included in Annex I. The greenhouse gas emissions by sources and removals by sinks associated with those activities shall be reported in a transparent and verifiable manner and reviewed in accordance with Articles 7 and 8.

4. Prior to the first session of the Conference of the Parties serving as the meeting of the Parties to this Protocol, each Party included in Annex I shall provide, for consideration by the Subsidiary Body for Scientific and Technological Advice, data to establish its level of carbon stocks in 1990 and to enable an estimate to be made of its changes in carbon stocks in subsequent years. The Conference of the Parties serving as the meeting of the Parties to this Protocol shall, at its first session or as soon as practicable thereafter, decide upon modalities, rules and guidelines as to how, and which, additional human-induced activities related to changes in greenhouse gas emissions by sources and removals by sinks in the agricultural soils and the land-use change and forestry categories shall be added to, or subtracted from, the assigned amounts for Parties included in Annex I, taking into account uncertainties, transparency in reporting, verifiability, the methodological work of the Intergovernmental Panel on Climate Change, the advice provided by the Subsidiary Body for Scientific and Technological Advice in accordance with Article 5 and the decisions of the Conference of the Parties. Such a decision shall apply in the second and subsequent commitment periods. A Party may choose to apply such a decision on

these additional human-induced activities for its first commitment period, provided that these activities have taken place since 1990.

5. The Parties included in Annex I undergoing the process of transition to a market economy whose base year or period was established pursuant to decision 9/CP.2 of the Conference of the Parties at its second session shall use that base year or period for the implementation of their commitments under this Article. Any other Party included in Annex I undergoing the process of transition to a market economy which has not yet submitted its first national communication under Article 12 of the Convention may also notify the Conference of the Parties serving as the meeting of the Parties to this Protocol that it intends to use an historical base year or period other than 1990 for the implementation of its commitments under this Article. The Conference of the Parties serving as the meeting of the Parties to this Protocol shall decide on the acceptance of such notification.

6. Taking into account Article 4, paragraph 6, of the Convention, in the implementation of their commitments under this Protocol other than those under this Article, a certain degree of flexibility shall be allowed by the Conference of the Parties serving as the meeting of the Parties to this Protocol to the Parties included in Annex I undergoing the process of transition to a market economy.

7. In the first quantified emission limitation and reduction commitment period, from 2008 to 2012, the assigned amount for each Party included in Annex I shall be equal to the percentage inscribed for it in Annex B of its aggregate anthropogenic carbon dioxide equivalent emissions of the greenhouse gases listed in Annex A in 1990, or the base

year or period determined in accordance with paragraph 5 above, multiplied by five. Those Parties included in Annex I for whom land-use change and forestry constituted a net source of greenhouse gas emissions in 1990 shall include in their 1990 emissions base year or period the aggregate anthropogenic carbon dioxide equivalent emissions by sources minus removals by sinks in 1990 from land-use change for the purposes of calculating their assigned amount.

8. Any Party included in Annex I may use 1995 as its base year for hydrofluorocarbons, perfluorocarbons and sulphur hexafluoride, for the purposes of the calculation referred to in paragraph 7 above.

9. Commitments for subsequent periods for Parties included in Annex I shall be established in amendments to Annex B to this Protocol, which shall be adopted in accordance with the provisions of Article 21, paragraph 7. The Conference of the Parties serving as the meeting of the Parties to this Protocol shall initiate the consideration of such commitments at least seven years before the end of the first commitment period referred to in paragraph 1 above.

10. Any emission reduction units, or any part of an assigned amount, which a Party acquires from another Party in accordance with the provisions of Article 6 or of Article 17 shall be added to the assigned amount for the acquiring Party.

11. Any emission reduction units, or any part of an assigned amount, which a Party transfers to another Party in accordance with the provisions of Article 6 or of Article 17 shall be subtracted from the assigned amount for the transferring Party.

12. Any certified emission reductions which a Party acquires from another Party in accordance with the provisions of Article 12 shall be added to the assigned amount for the acquiring Party.

13. If the emissions of a Party included in Annex I in a commitment period are less than its assigned amount under this Article, this difference shall, on request of that Party, be added to the assigned amount for that Party for subsequent commitment periods.

14. Each Party included in Annex I shall strive to implement the commitments mentioned in paragraph 1 above in such a way as to minimize adverse social, environmental and economic impacts on developing country Parties, particularly those identified in Article 4, paragraphs 8 and 9, of the Convention. In line with relevant decisions of the Conference of the Parties on the implementation of those paragraphs, the Conference of the Parties serving as the meeting of the Parties to this Protocol shall, at its first session, consider what actions are necessary to minimize the adverse effects of climate change and/or the impacts of response measures on Parties referred to in those paragraphs. Among the issues to be considered shall be the establishment of funding, insurance and transfer of technology.

Article 4

1. Any Parties included in Annex I that have reached an agreement to fulfil their commitments under Article 3 jointly, shall be deemed to have met those commitments provided that their total combined aggregate anthropogenic carbon dioxide equivalent emissions of the greenhouse gases listed in Annex A do not exceed their assigned amounts

calculated pursuant to their quantified emission limitation and reduction commitments inscribed in Annex B and in accordance with the provisions of Article 3. The respective emission level allocated to each of the Parties to the agreement shall be set out in that agreement.

2. The Parties to any such agreement shall notify the secretariat of the terms of the agreement on the date of deposit of their instruments of ratification, acceptance or approval of this Protocol, or accession thereto. The secretariat shall in turn inform the Parties and signatories to the Convention of the terms of the agreement.

3. Any such agreement shall remain in operation for the duration of the commitment period specified in Article 3, paragraph 7.

4. If Parties acting jointly do so in the framework of, and together with, a regional economic integration organization, any alteration in the composition of the organization after adoption of this Protocol shall not affect existing commitments under this Protocol. Any alteration in the composition of the organization shall only apply for the purposes of those commitments under Article 3 that are adopted subsequent to that alteration.

5. In the event of failure by the Parties to such an agreement to achieve their total combined level of emission reductions, each Party to that agreement shall be responsible for its own level of emissions set out in the agreement.

6. If Parties acting jointly do so in the framework of, and together with, a regional economic integration organization which is itself a Party to this Protocol, each member State of that regional economic integration organization individually, and together with the regional

economic integration organization acting in accordance with Article 24, shall, in the event of failure to achieve the total combined level of emission reductions, be responsible for its level of emissions as notified in accordance with this Article.

Article 5

1. Each Party included in Annex I shall have in place, no later than one year prior to the start of the first commitment period, a national system for the estimation of anthropogenic emissions by sources and removals by sinks of all greenhouse gases not controlled by the Montreal Protocol. Guidelines for such national systems, which shall incorporate the methodologies specified in paragraph 2 below, shall be decided upon by the Conference of the Parties serving as the meeting of the Parties to this Protocol at its first session.

2. Methodologies for estimating anthropogenic emissions by sources and removals by sinks of all greenhouse gases not controlled by the Montreal Protocol shall be those accepted by the Intergovernmental Panel on Climate Change and agreed upon by the Conference of the Parties at its third session. Where such methodologies are not used, appropriate adjustments shall be applied according to methodologies agreed upon by the Conference of the Parties serving as the meeting of the Parties to this Protocol at its first session. Based on the work of, *inter alia*, the Intergovernmental Panel on Climate Change and advice provided by the Subsidiary Body for Scientific and Technological Advice, the Conference of the Parties serving as the meeting of the Parties to this Protocol shall regularly review and, as appropriate, revise such methodologies and adjustments, taking fully into account any relevant decisions by the Conference of the Parties. Any revision to

methodologies or adjustments shall be used only for the purposes of ascertaining compliance with commitments under Article 3 in respect of any commitment period adopted subsequent to that revision.

3. The global warming potentials used to calculate the carbon dioxide equivalence of anthropogenic emissions by sources and removals by sinks of greenhouse gases listed in Annex A shall be those accepted by the Intergovernmental Panel on Climate Change and agreed upon by the Conference of the Parties at its third session. Based on the work of, *inter alia*, the Intergovernmental Panel on Climate Change and advice provided by the Subsidiary Body for Scientific and Technological Advice, the Conference of the Parties serving as the meeting of the Parties to this Protocol shall regularly review and, as appropriate, revise the global warming potential of each such greenhouse gas, taking fully into account any relevant decisions by the Conference of the Parties. Any revision to a global warming potential shall apply only to commitments under Article 3 in respect of any commitment period adopted subsequent to that revision.

Article 6

1. For the purpose of meeting its commitments under Article 3, any Party included in Annex I may transfer to, or acquire from, any other such Party emission reduction units resulting from projects aimed at reducing anthropogenic emissions by sources or enhancing anthropogenic removals by sinks of greenhouse gases in any sector of the economy, provided that:

(a) Any such project has the approval of the Parties involved;

(b) Any such project provides a reduction in emissions by sources, or an enhancement of removals by sinks, that is additional to any that would otherwise occur;

(c) It does not acquire any emission reduction units if it is not in compliance with its obligations under Articles 5 and 7; and

(d) The acquisition of emission reduction units shall be supplemental to domestic actions for the purposes of meeting commitments under Article 3.

2. The Conference of the Parties serving as the meeting of the Parties to this Protocol may, at its first session or as soon as practicable thereafter, further elaborate guidelines for the implementation of this Article, including for verification and reporting.

3. A Party included in Annex I may authorize legal entities to participate, under its responsibility, in actions leading to the generation, transfer or acquisition under this Article of emission reduction units.

4. If a question of implementation by a Party included in Annex I of the requirements referred to in this Article is identified in accordance with the relevant provisions of Article 8, transfers and acquisitions of emission reduction units may continue to be made after the question has been identified, provided that any such units may not be used by a Party to meet its commitments under Article 3 until any issue of compliance is resolved.

Article 7

1. Each Party included in Annex I shall incorporate in

its annual inventory of anthropogenic emissions by sources and removals by sinks of greenhouse gases not controlled by the Montreal Protocol, submitted in accordance with the relevant decisions of the Conference of the Parties, the necessary supplementary information for the purposes of ensuring compliance with Article 3, to be determined in accordance with paragraph 4 below.

2. Each Party included in Annex I shall incorporate in its national communication, submitted under Article 12 of the Convention, the supplementary information necessary to demonstrate compliance with its commitments under this Protocol, to be determined in accordance with paragraph 4 below.

3. Each Party included in Annex I shall submit the information required under paragraph 1 above annually, beginning with the first inventory due under the Convention for the first year of the commitment period after this Protocol has entered into force for that Party. Each such Party shall submit the information required under paragraph 2 above as part of the first national communication due under the Convention after this Protocol has entered into force for it and after the adoption of guidelines as provided for in paragraph 4 below. The frequency of subsequent submission of information required under this Article shall be determined by the Conference of the Parties serving as the meeting of the Parties to this Protocol, taking into account any timetable for the submission of national communications decided upon by the Conference of the Parties.

4. The Conference of the Parties serving as the meeting of the Parties to this Protocol shall adopt at its first session, and review periodically thereafter, guidelines for the

preparation of the information required under this Article, taking into account guidelines for the preparation of national communications by Parties included in Annex I adopted by the Conference of the Parties. The Conference of the Parties serving as the meeting of the Parties to this Protocol shall also, prior to the first commitment period, decide upon modalities for the accounting of assigned amounts.

Article 8

1. The information submitted under Article 7 by each Party included in Annex I shall be reviewed by expert review teams pursuant to the relevant decisions of the Conference of the Parties and in accordance with guidelines adopted for this purpose by the Conference of the Parties serving as the meeting of the Parties to this Protocol under paragraph 4 below. The information submitted under Article 7, paragraph 1, by each Party included in Annex I shall be reviewed as part of the annual compilation and accounting of emissions inventories and assigned amounts. Additionally, the information submitted under Article 7, paragraph 2, by each Party included in Annex I shall be reviewed as part of the review of communications.

2. Expert review teams shall be coordinated by the secretariat and shall be composed of experts selected from those nominated by Parties to the Convention and, as appropriate, by intergovernmental organizations, in accordance with guidance provided for this purpose by the Conference of the Parties.

3. The review process shall provide a thorough and comprehensive technical assessment of all aspects of the implementation by a Party of this Protocol. The expert review teams shall prepare a report to the Conference of the Parties serving as the meeting of the Parties to this

Protocol, assessing the implementation of the commitments of the Party and identifying any potential problems in, and factors influencing, the fulfilment of commitments. Such reports shall be circulated by the secretariat to all Parties to the Convention. The secretariat shall list those questions of implementation indicated in such reports for further consideration by the Conference of the Parties serving as the meeting of the Parties to this Protocol.

4. The Conference of the Parties serving as the meeting of the Parties to this Protocol shall adopt at its first session, and review periodically thereafter, guidelines for the review of implementation of this Protocol by expert review teams taking into account the relevant decisions of the Conference of the Parties.

5. The Conference of the Parties serving as the meeting of the Parties to this Protocol shall, with the assistance of the Subsidiary Body for Implementation and, as appropriate, the Subsidiary Body for Scientific and Technological Advice, consider:

(a) The information submitted by Parties under Article 7 and the reports of the expert reviews thereon conducted under this Article; and

(b) Those questions of implementation listed by the secretariat under paragraph 3 above, as well as any questions raised by Parties.

6. Pursuant to its consideration of the information referred to in paragraph 5 above, the Conference of the Parties serving as the meeting of the Parties to this Protocol shall take decisions on any matter required for the implementation of this Protocol.

Article 9

1. The Conference of the Parties serving as the meeting of the Parties to this Protocol shall periodically review this Protocol in the light of the best available scientific information and assessments on climate change and its impacts, as well as relevant technical, social and economic information. Such reviews shall be coordinated with pertinent reviews under the Convention, in particular those required by Article 4, paragraph 2(d), and Article 7, paragraph 2(a), of the Convention. Based on these reviews, the Conference of the Parties serving as the meeting of the Parties to this Protocol shall take appropriate action.

2. The first review shall take place at the second session of the Conference of the Parties serving as the meeting of the Parties to this Protocol. Further reviews shall take place at regular intervals and in a timely manner.

Article 10

All Parties, taking into account their common but differentiated responsibilities and their specific national and regional development priorities, objectives and circumstances, without introducing any new commitments for Parties not included in Annex I, but reaffirming existing commitments under Article 4, paragraph 1, of the Convention, and continuing to advance the implementation of these commitments in order to achieve sustainable development, taking into account Article 4, paragraphs 3, 5 and 7, of the Convention, shall:

(a) Formulate, where relevant and to the extent possible, cost-effective national and, where appropriate, regional programmes to improve the quality of local emission factors, activity data and/or models which reflect the socio-economic conditions of each Party for the preparation and periodic updating of national inventories of anthropogenic emissions by sources and removals by sinks of all greenhouse gases not controlled by the Montreal Protocol, using comparable methodologies to be agreed upon by the Conference of the Parties, and consistent with the guidelines for the preparation of national communications adopted by the Conference of the Parties;

(b) Formulate, implement, publish and regularly update national and, where appropriate, regional programmes containing measures to mitigate climate change and measures to facilitate adequate adaptation to climate change:

(i) Such programmes would, *inter alia*, concern the energy, transport and industry sectors as well as agriculture, forestry and waste management. Furthermore, adaptation technologies and methods for improving spatial planning would improve adaptation to climate change; and

(ii) Parties included in Annex I shall submit information on action under this Protocol, including national programmes, in accordance with Article 7; and other Parties shall seek to include in their national communications, as appropriate, information on programmes which contain measures that the Party believes contribute to addressing climate change and its adverse impacts, including the abatement of increases in greenhouse gas emissions, and enhancement of and removals by sinks, capacity building and adaptation measures;

(c) Cooperate in the promotion of effective modalities for the development, application and diffusion of, and take all practicable steps to promote, facilitate and finance, as appropriate, the transfer of, or access to, environmentally sound technologies, know-how, practices and processes pertinent to climate change, in particular to developing countries, including the formulation of policies and programmes for the effective transfer of environmentally sound technologies that are publicly owned or in the public domain and the creation of an enabling environment for the private sector, to promote and enhance the transfer of, and access to, environmentally sound technologies;

(d) Cooperate in scientific and technical research and promote the maintenance and the development of systematic observation systems and development of data archives to reduce uncertainties related to the climate system, the adverse impacts of climate change and the economic and social consequences of various response strategies, and promote the development and strengthening of endogenous capacities and capabilities to participate in international and intergovernmental efforts, programmes and networks on research and systematic observation, taking into account Article 5 of the Convention;

(e) Cooperate in and promote at the international level, and, where appropriate, using existing bodies, the development and implementation of education and training programmes, including the strengthening of national capacity building, in particular human and institutional capacities and the exchange or secondment of personnel to train experts in this field, in particular for developing countries, and facilitate at the national level public awareness of, and public access to information on, climate change. Suitable modalities should be developed to implement these activities through the relevant bodies of the Convention, taking into account Article 6 of the Convention;

(f) Include in their national communications information on programmes and activities undertaken pursuant to this Article in accordance with relevant decisions of the Conference of the Parties; and

(g) Give full consideration, in implementing the commitments under this Article, to Article 4, paragraph 8, of the Convention.

Article 11

1. In the implementation of Article 10, Parties shall take into account the provisions of Article 4, paragraphs 4, 5, 7, 8 and 9, of the Convention.

2. In the context of the implementation of Article 4, paragraph 1, of the Convention, in accordance with the provisions of Article 4, paragraph 3, and Article 11 of the Convention, and through the entity or entities entrusted with the operation of the financial mechanism of the Convention, the developed country Parties and other developed Parties included in Annex II to the Convention shall:

(a) Provide new and additional financial resources to meet the agreed full costs incurred by developing country Parties in advancing the implementation of existing commitments under Article 4, paragraph 1(a), of the Convention that are covered in Article 10, subparagraph (a); and

(b) Also provide such financial resources, including for the transfer of technology, needed by the developing country Parties to meet the agreed full incremental costs of advancing the implementation of existing commitments under Article 4, paragraph 1, of the Convention that are covered by Article 10 and that are agreed between a developing country Party and the international entity or entities referred to in Article 11 of the Convention, in accordance with that Article.

The implementation of these existing commitments shall take into account the need for adequacy and predictability in the flow of funds and the importance of appropriate burden sharing among developed country Parties. The guidance to the entity or entities entrusted with the operation of the financial mechanism of the Convention in relevant decisions of the Conference of the Parties, including those agreed before the adoption of this Protocol, shall apply *mutatis mutandis* to the provisions of this paragraph.

3. The developed country Parties and other developed Parties in Annex II to the Convention may also provide, and developing country Parties avail themselves of, financial resources for the implementation of Article 10, through bilateral, regional and other multilateral channels.

Article 12

1. A clean development mechanism is hereby defined.

2. The purpose of the clean development mechanism shall be to assist Parties not included in Annex I in achieving sustainable development and in contributing to the ultimate objective of the Convention, and to assist Parties included in Annex I in achieving compliance with their quantified emission limitation and reduction commitments under Article 3.

3. Under the clean development mechanism:

(a) Parties not included in Annex I will benefit from project activities resulting in certified emission reductions; and

(b) Parties included in Annex I may use the certified emission reductions accruing from such project activities to contribute to compliance with part of their quantified emission limitation and reduction commitments under Article 3, as determined by the Conference of the Parties serving as the meeting of the Parties to this Protocol.

4. The clean development mechanism shall be subject to the authority and guidance of the Conference of the Parties serving as the meeting of the Parties to this Protocol and be supervised by an executive board of the clean development mechanism.

5. Emission reductions resulting from each project activity shall be certified by operational entities to be designated by the Conference of the Parties serving as the meeting of the Parties to this Protocol, on the basis of:

(a) Voluntary participation approved by each Party involved;

(b) Real, measurable, and long-term benefits related to the mitigation of climate change; and

(c) Reductions in emissions that are additional to any that would occur in the absence of the certified project activity.

6. The clean development mechanism shall assist in arranging funding of certified project activities as necessary.

7. The Conference of the Parties serving as the meeting of the Parties to this Protocol shall, at its first session, elaborate modalities and procedures with the objective of ensuring transparency, efficiency and accountability through independent auditing and verification of project activities.

8. The Conference of the Parties serving as the meeting of the Parties to this Protocol shall ensure that a share of the proceeds from certified project activities is used to cover administrative expenses as well as to assist developing country Parties that are particularly vulnerable to the adverse effects of climate change to meet the costs of adaptation.

9. Participation under the clean development mechanism, including in activities mentioned in paragraph 3(a) above and in the acquisition of certified emission reductions, may involve private and/or public entities, and is to be subject to whatever guidance may be provided by the executive board of the clean development mechanism.

10. Certified emission reductions obtained during the period from the year 2000 up to the beginning of the first commitment period can be used to assist in achieving compliance in the first commitment period.

Article 13

1. The Conference of the Parties, the supreme body of the Convention, shall serve as the meeting of the Parties to this Protocol.

2. Parties to the Convention that are not Parties to this Protocol may participate as observers in the proceedings of any session of the Conference of the Parties serving as the meeting of the Parties to this Protocol. When the Conference of the Parties serves as the meeting of the Parties to this Protocol, decisions under this Protocol shall be taken only by those that are Parties to this Protocol.

3. When the Conference of the Parties serves as the meeting of the Parties to this Protocol, any member of the Bureau of the Conference of the Parties representing a Party to the Convention but, at that time, not a Party to this Protocol, shall be replaced by an additional member to be elected by and from amongst the Parties to this Protocol.

4. The Conference of the Parties serving as the meeting of the Parties to this Protocol shall keep under regular review the implementation of this Protocol and shall make, within its mandate, the decisions necessary to promote its effective implementation. It shall perform the functions assigned to it by this Protocol and shall:

(a) Assess, on the basis of all information made available to it in accordance with the provisions of this Protocol, the implementation of this Protocol by the Parties, the overall effects of the measures taken pursuant to this Protocol, in particular environmental, economic and social effects as well as their cumulative impacts and the extent to which progress towards the objective of the Convention is being achieved;

(b) Periodically examine the obligations of the Parties under this Protocol, giving due consideration to any reviews required by Article 4, paragraph 2(d), and Article 7, paragraph 2, of the Convention, in the light of the objective of the Convention, the experience gained in its implementation and the evolution of scientific and technological knowledge, and in this respect consider and adopt regular reports on the implementation of this Protocol;

(c) Promote and facilitate the exchange of information on measures adopted by the Parties to address climate change and its effects, taking into account the differing circumstances, responsibilities and capabilities of the Parties and their respective commitments under this Protocol;

(d) Facilitate, at the request of two or more Parties, the coordination of measures adopted by them to address climate change and its effects, taking into account the differing circumstances, responsibilities and capabilities of the Parties and their respective commitments under this Protocol;

(e) Promote and guide, in accordance with the objective of the Convention and the provisions of this Protocol, and taking fully into account the relevant decisions by the Conference of the Parties, the development and periodic refinement of comparable methodologies for the effective implementation of this Protocol, to be agreed on by the Conference of the Parties serving as the meeting of the Parties to this Protocol;

(f) Make recommendations on any matters necessary for the implementation of this Protocol;

(g) Seek to mobilize additional financial resources in accordance with Article 11, paragraph 2;

(h) Establish such subsidiary bodies as are deemed necessary for the implementation of this Protocol;

(i) Seek and utilize, where appropriate, the services and cooperation of, and information provided by, competent international organizations and intergovernmental and non-governmental bodies; and

(j) Exercise such other functions as may be required for the implementation of this Protocol, and consider any assignment resulting from a decision by the Conference of the Parties.

5. The rules of procedure of the Conference of the Parties and financial procedures applied under the Convention shall be applied *mutatis mutandis* under this Protocol, except as may be otherwise decided by consensus by the Conference of the Parties serving as the meeting of the Parties to this Protocol.

6. The first session of the Conference of the Parties serving as the meeting of the Parties to this Protocol shall be convened by the secretariat in conjunction with the first session of the Conference of the Parties that is scheduled after the date of the entry into force of this Protocol. Subsequent ordinary sessions of the Conference of the Parties serving as the meeting of the Parties to this Protocol shall be held every year and in conjunction with ordinary sessions of the Conference of the Parties, unless otherwise decided by the Conference of the Parties serving as the meeting of the Parties to this Protocol.

7. Extraordinary sessions of the Conference of the Parties

serving as the meeting of the Parties to this Protocol shall be held at such other times as may be deemed necessary by the Conference of the Parties serving as the meeting of the Parties to this Protocol, or at the written request of any Party, provided that, within six months of the request being communicated to the Parties by the secretariat, it is supported by at least one third of the Parties.

8. The United Nations, its specialized agencies and the International Atomic Energy

Agency, as well as any State member thereof or observers thereto not party to the Convention, may be represented at sessions of the Conference of the Parties serving as the meeting of the Parties to this Protocol as observers. Any body or agency, whether national or international, governmental or non-governmental, which is qualified in matters covered by this Protocol and which has informed the secretariat of its wish to be represented at a session of the Conference of the Parties serving as the meeting of the Parties to this Protocol as an observer, may be so admitted unless at least one third of the Parties present object. The admission and participation of observers shall be subject to the rules of procedure, as referred to in paragraph 5 above.

Article 14

1. The secretariat established by Article 8 of the Convention shall serve as the secretariat of this Protocol.

2. Article 8, paragraph 2, of the Convention on the functions of the secretariat, and

Article 8, paragraph 3, of the Convention on arrangements made for the functioning of the secretariat, shall apply *mutatis mutandis* to this Protocol. The secretariat shall, in addition, exercise the functions assigned to it under this Protocol.

Article 15

1. The Subsidiary Body for Scientific and Technological Advice and the Subsidiary Body for Implementation established by Articles 9 and 10 of the Convention shall serve as, respectively, the Subsidiary Body for Scientific and Technological Advice and the Subsidiary Body for Implementation of this Protocol. The provisions relating to the functioning of these two bodies under the Convention shall apply *mutatis mutandis* to this Protocol. Sessions of the meetings of the Subsidiary Body for Scientific and Technological Advice and the Subsidiary Body for Implementation of this Protocol shall be held in conjunction with the meetings of, respectively, the Subsidiary Body for Scientific and Technological Advice and the Subsidiary Body for Implementation of the Convention.

2. Parties to the Convention that are not Parties to this Protocol may participate as observers in the proceedings of any session of the subsidiary bodies. When the subsidiary bodies serve as the subsidiary bodies of this Protocol, decisions under this Protocol shall be taken only by those that are Parties to this Protocol.

3. When the subsidiary bodies established by Articles 9 and 10 of the Convention exercise their functions with regard to matters concerning this Protocol, any member of the Bureaux of those subsidiary bodies representing a Party to the Convention but, at that time, not a party to this Protocol, shall be replaced by an additional member to be elected by and from amongst the Parties to this Protocol.

Article 16

The Conference of the Parties serving as the meeting of the Parties to this Protocol shall, as soon as practicable, consider the application to this Protocol of, and modify as appropriate, the multilateral consultative process referred to in Article 13 of the Convention, in the light of any relevant decisions that may be taken by the Conference of the Parties. Any multilateral consultative process that may be applied to this Protocol shall operate without prejudice to the procedures and mechanisms established in accordance with Article 18.

Article 17

The Conference of the Parties shall define the relevant principles, modalities, rules and guidelines, in particular for verification, reporting and accountability for emissions trading. The Parties included in Annex B may participate in emissions trading for the purposes of fulfilling their commitments under Article 3. Any such trading shall be supplemental to domestic actions for the purpose of meeting quantified emission limitation and reduction commitments under that Article.

Article 18

The Conference of the Parties serving as the meeting of the Parties to this Protocol shall, at its first session, approve appropriate and effective procedures and mechanisms to determine and to address cases of non-compliance with the provisions of this Protocol, including through the development of an indicative list of consequences, taking into account the cause, type, degree and frequency of non-compliance. Any procedures and mechanisms under this Article entailing binding consequences shall be adopted by means of an amendment to this Protocol.

Article 19

The provisions of Article 14 of the Convention on settlement of disputes shall apply *mutatis mutandis* to this Protocol.

Article 20

1. Any Party may propose amendments to this Protocol.

2. Amendments to this Protocol shall be adopted at an ordinary session of the Conference of the Parties serving as the meeting of the Parties to this Protocol. The text of any proposed amendment to this Protocol shall be communicated to the Parties by the secretariat at least six months before the meeting at which it is proposed for adoption. The secretariat shall also communicate the text of any proposed amendments to the Parties and signatories to the Convention and, for information, to the Depositary.

3. The Parties shall make every effort to reach agreement on any proposed amendment to this Protocol by consensus. If all efforts at consensus have been exhausted, and no agreement reached, the amendment shall as a last resort be adopted by a three-fourths majority vote of the Parties present and voting at the meeting. The adopted amendment shall be communicated by the secretariat to the Depositary, who shall circulate it to all Parties for their acceptance.

4. Instruments of acceptance in respect of an amendment shall be deposited with the Depositary. An amendment adopted in accordance with paragraph 3 above shall enter into force for those Parties having accepted it on the ninetieth day after the date of receipt by the Depositary of an instrument of acceptance by at least three fourths of the Parties to this Protocol.

5. The amendment shall enter into force for any other Party on the ninetieth day after the date on which that Party deposits with the Depositary its instrument of acceptance of the said amendment.

Article 21

1. Annexes to this Protocol shall form an integral part thereof and, unless otherwise expressly provided, a reference to this Protocol constitutes at the same time a reference to any annexes thereto. Any annexes adopted after the entry into force of this Protocol shall be restricted to lists, forms and any other material of a descriptive nature that is of a scientific, technical, procedural or administrative character.

2. Any Party may make proposals for an annex to this Protocol and may propose amendments to annexes to this Protocol.

3. Annexes to this Protocol and amendments to annexes to this Protocol shall be adopted at an ordinary session of the Conference of the Parties serving as the meeting of the Parties to this Protocol. The text of any proposed annex or amendment to an annex shall be communicated to the Parties by the secretariat at least six months before the meeting at which it is proposed for adoption. The secretariat shall also communicate the text of any proposed annex or amendment to an annex to the Parties and signatories to the Convention and, for information, to the Depositary.

4. The Parties shall make every effort to reach agreement on any proposed annex or amendment to an annex by consensus. If all efforts at consensus have been exhausted, and no agreement reached, the annex or amendment to an annex shall as a last resort be adopted by a three-fourths majority vote of the Parties present and voting at the

meeting. The adopted annex or amendment to an annex shall be communicated by the secretariat to the Depositary, who shall circulate it to all Parties for their acceptance.

5. An annex, or amendment to an annex other than Annex A or B, that has been adopted in accordance with paragraphs 3 and 4 above shall enter into force for all Parties to this Protocol six months after the date of the communication by the Depositary to such Parties of the adoption of the annex or adoption of the amendment to the annex, except for those Parties that have notified the Depositary, in writing, within that period of their non-acceptance of the annex or amendment to the annex. The annex or amendment to an annex shall enter into force for Parties which withdraw their notification of non-acceptance on the ninetieth day after the date on which withdrawal of such notification has been received by the Depositary.

6. If the adoption of an annex or an amendment to an annex involves an amendment to this Protocol, that annex or amendment to an annex shall not enter into force until such time as the amendment to this Protocol enters into force.

7. Amendments to Annexes A and B to this Protocol shall be adopted and enter into force in accordance with the procedure set out in Article 20, provided that any amendment to Annex B shall be adopted only with the written consent of the Party concerned.

Article 22

1. Each Party shall have one vote, except as provided for in paragraph 2 below.

2. Regional economic integration organizations, in matters within their competence, shall exercise their right to vote

with a number of votes equal to the number of their member States that are Parties to this Protocol. Such an organization shall not exercise its right to vote if any of its member States exercises its right, and vice versa.

Article 23

The Secretary-General of the United Nations shall be the Depositary of this Protocol.

Article 24

1. This Protocol shall be open for signature and subject to ratification, acceptance or approval by States and regional economic integration organizations which are Parties to the Convention. It shall be open for signature at United Nations Headquarters in New York from

16 March 1998 to 15 March 1999. This Protocol shall be open for accession from the day after the date on which it is closed for signature. Instruments of ratification, acceptance, approval or accession shall be deposited with the Depositary.

2. Any regional economic integration organization which becomes a Party to this Protocol without any of its member States being a Party shall be bound by all the obligations under this Protocol. In the case of such organizations, one or more of whose member States is a Party to this Protocol, the organization and its member States shall decide on their respective responsibilities for the performance of their obligations under this Protocol. In such cases, the organization and the member States shall not be entitled to exercise rights under this Protocol concurrently.

3. In their instruments of ratification, acceptance, approval or accession, regional economic integration

organizations shall declare the extent of their competence with respect to the matters governed by this Protocol. These organizations shall also inform the Depositary, who shall in turn inform the Parties, of any substantial modification in the extent of their competence.

Article 25

1. This Protocol shall enter into force on the ninetieth day after the date on which not less than 55 Parties to the Convention, incorporating Parties included in Annex I which accounted in total for at least 55 per cent of the total carbon dioxide emissions for 1990 of the Parties included in Annex I, have deposited their instruments of ratification, acceptance, approval or accession.

2. For the purposes of this Article, "the total carbon dioxide emissions for 1990 of the Parties included in Annex I" means the amount communicated on or before the date of adoption of this Protocol by the Parties included in Annex I in their first national communications submitted in accordance with Article 12 of the Convention.

3. For each State or regional economic integration organization that ratifies, accepts or approves this Protocol or accedes thereto after the conditions set out in paragraph 1 above for entry into force have been fulfilled, this Protocol shall enter into force on the ninetieth day following the date of deposit of its instrument of ratification, acceptance, approval or accession.

4. For the purposes of this Article, any instrument deposited by a regional economic integration organization shall not be counted as additional to those deposited by States members of the organization.

Article 26

No reservations may be made to this Protocol.

Article 27

1. At any time after three years from the date on which this Protocol has entered into force for a Party, that Party may withdraw from this Protocol by giving written notification to the Depositary.

2. Any such withdrawal shall take effect upon expiry of one year from the date of receipt by the Depositary of the notification of withdrawal, or on such later date as may be specified in the notification of withdrawal.

3. Any Party that withdraws from the Convention shall be considered as also having withdrawn from this Protocol.

Article 28

The original of this Protocol, of which the Arabic, Chinese, English, French, Russian and Spanish texts are equally authentic, shall be deposited with the Secretary-General of the United Nations.

Done at Kyoto this eleventh day of December one thousand nine hundred and ninety-seven.

In Witness Whereof the undersigned, being duly authorized to that effect, have affixed their signatures to this Protocol on the dates indicated.

The Kyoto Protocol: An Overview

This international agreement, which builds on the United Nations Framework Convention on Climate Change, sets legally binding targets and timetables for cutting the greenhouse-gas emissions of industrialized countries. The

Protocol, with its stiffer requirements, will have a real impact on climate change. Under the Protocol, countries may buy and sell greenhouse-gas emissions "units" and "credits." The Protocol provides a system for financing emissions-reducing or emissions-avoiding projects in developing nations. Within the Protocol, industrialized countries are granted "emissions reduction units" for financing projects in other developed countries — a system likely to increase efficiency and reduce the global-warming output of the "transition economies" of central and eastern Europe.

It took all of one year for the member countries of the Framework Convention on Climate Change to decide that the Convention had to be augmented by an agreement with stricter demands for reducing greenhouse-gas emissions. The Convention took effect in 1994, and by 1995 governments had begun negotiations on a protocol — an international agreement linked to the existing treaty, but standing on its own. The text of the Kyoto Protocol was adopted unanimously in 1997; it entered into force on 16 February 2005. The Protocol's major feature is that it has mandatory targets on greenhouse-gas emissions for the world's leading economies which have accepted it. These targets range from -8 per cent to +10 per cent of the countries' individual 1990 emissions levels "with a view to reducing their overall emissions of such gases by at least 5 per cent below existing 1990 levels in the commitment period 2008 to 2012." In almost all cases — even those set at +10 per cent of 1990 levels — the limits call for significant reductions in currently projected emissions. Future mandatory targets are expected to be established for "commitment periods" after 2012. These are to be negotiated well in advance of the periods concerned.

Commitments under the Protocol vary from nation to nation. The overall 5 per cent target for developed countries is to be met through cuts (from 1990 levels) of 8 per cent in the European Union (EU[15]), Switzerland, and most Central and East European states; 6 per cent in Canada; 7 per cent in the United States (although the US has since withdrawn its support for the Protocol); and 6 per cent in Hungary, Japan, and Poland. New Zealand, Russia, and Ukraine are to stabilize their emissions, while Norway may increase emissions by up to 1 per cent, Australia by up to 8 per cent (subsequently withdrew its support for the Protocol), and Iceland by 10 per cent. The EU has made its own internal agrement to meet its 8 per cent target by distributing different rates to its member states. These targets range from a 28 per cent reduction by Luxembourg and 21 per cent cuts by Denmark and Germany to a 25 per cent increase by Greece and a 27 per cent increase by Portugal. To compensate for the sting of "binding targets," as they are called, the agreement offers flexibility in how countries may meet their targets. For example, they may partially compensate for their emissions by increasing "sinks" — forests, which remove carbon dioxide from the atmosphere. That may be accomplished either on their own territories or in other countries. Or they may pay for foreign projects that result in greenhouse-gas cuts. Several mechanisms have been set up for this purpose.

The Kyoto Protocol is a complicated agreement that has been slow in coming—there are reasons for this. The Protocol not only has to be an effective against a complicated worldwide problem — it also has to be politically acceptable. As a result, panels and committees have multiplied to monitor and referee its various programmes, and even after the agreement was approved in 1997, further negotiations were deemed necessary to hammer out instructions on how to "operate" it. These rules, adopted in 2001, are called the "Marrakesh Accords." There is a delicate balance to international treaties. Those appealing enough to gain widespread support often aren't strong enough to solve the problems they focus on. (Because the Framework Convention was judged to have this weakness, despite its many valuable provisions, the Protocol was created to supplement it.) Yet treaties with real "teeth" may have difficulty attracting enough widespread support to be effective. Some mechanisms of the Protocol had enough support that they were set up in advance of the Protocol's entry into force. The Clean Development Mechanism, for example — through which industrialized countries can partly meet their binding emissions targets through "credits" earned by sponsoring greenhouse-gas-reducing projects in developing countries — already had an executive board before the Kyoto Protocol entered into force on 16 February 2005.

The limits on greenhouse-gas emissions set by the Kyoto Protocol are a way of assigning monetary value to the earth's shared atmosphere — something that has been missing up to now. Nations that have contributed the most to global warming have tended to benefit directly in terms of greater business profits and higher standards of living, while they have not been held proportionately accountable for the damages caused by their emissions. The negative effects of climate change will be felt all over the world, and actually the consequences are expected to be most severe in least-developed nations which have produced few emissions. The Kyoto Protocol sets limits on total emissions by the world's major economies, a prescribed number of "emission units." Individual industrialized countries will have mandatory emissions targets they must meet. . . but it is understood that some will do better than expected, coming in under their limits, while others will exceed them.

The Protocol allows countries that have emissions units to spare — emissions permitted them but not "used" — to sell this excess capacity to countries that are over their targets. This so-called "carbon market" — so-named because carbon dioxide is the most widely produced greenhouse gas, and because emissions of other greenhouse gases will be recorded and counted in terms of their "carbon dioxide equivalents" — is both flexible and realistic. Countries not meeting their commitments will be able to "buy" compliance. . . but the price may be steep.

The higher the cost, the more pressure they will feel to use energy more efficiently and to research and promote the development of alternative sources of energy that have low or no emissions. A global "stock market" where emissions units are bought and sold is simple in concept — but in practice the Protocol's emissions-trading system has been complicated to set up. The details, weren't specified in the Protocol, and so additional negotiations were held to hammer them out.

These rules were among the workaday specifics included in the 2001 "Marrakesh Accords." The problems are clear: countries' actual emissions have to be monitored and guaranteed to be what they are reported to be; and precise records have to be kept of the trades carried out. Accordingly, "registries"— like bank accounts of a nation's emissions units — are being set up, along with "accounting procedures," an "international transactions log," and "expert review teams" to police compliance.

More than actual emissions units will be involved in trades and sales. Countries will get credit for reducing greenhouse-gas totals by planting or expanding forests ("removal units"); for carrying out "joint implementation projects" with other developed countries, usually countries with "transition economies"; and for projects under the Protocol's Clean Development Mechanism, which involves funding activities to reduce emissions by developing nations.

Credits earned this way may be bought and sold in the emissions market or "banked" for future use. Some national registry systems under the Protocol have already been set up, as countries are eager to "bank" emissions reductions already accomplished while they wait for the Protocol to win its final ratifications and become legally binding. Smaller "carbon markets" were established by the European Union and other groups of countries; they were operating before the Protocol entered into force. These emissions-trading systems were intended to start the process and to link up with the Protocol's global market once it becomes operational.

The text of the Protocol to the UNFCCC was adopted at the third session of the Conference of the Parties to the UNFCCC in Kyoto, Japan, on 11 December 1997; it was open for signature from 16 March 1998 to 15 March 1999 at United Nations Headquarters, New York. By that date the Protocol had received 84 signatures. Those Parties that have not yet signed the Kyoto Protocol may accede to it at any time. The Protocol is subject to ratification, acceptance, approval or accession by Parties to the Convention.

It entered into force on 16 February 2005 — the ninetieth day after at least 55 Parties to the Convention, incorporating Annex I Parties which accounted in total for at least 55 %

of the total carbon dioxide emissions for 1990 from that group, deposited their instruments of ratification, acceptance, approval or accession. The list, to the right, contains the latest information concerning dates of signature and ratification received from the Secretary-General of the United Nations, as Depository of the Kyoto Protocol. The dates in the column entitled "date of ratification" are those of the receipt of the instrument of ratification (R), acceptance (At), approval (Ap) or accession (Ac).

ANNEX A

Greenhouse gases

Carbon dioxide (CO_2)

Methane (CH4)

Nitrous oxide (N2O)

Hydrofluorocarbons (HFCs)

Perfluorocarbons (PFCs)

Sulphur hexafluoride (SF6)

Sectors/source categories

Energy

Fuel combustion

Energy industries

Manufacturing industries and construction

Transport

Other sectors

Other

Fugitive emissions from fuels

Solid fuels

Oil and natural gas

Other

Industrial processes

Mineral products

Chemical industry

Metal production

Other production

Production of halocarbons and sulphur hexafluoride

Consumption of halocarbons and sulphur hexafluoride

Other

Solvent and other product use

Agriculture

Enteric fermentation

Manure management

Rice cultivation

Agricultural soils

Prescribed burning of savannas

Field burning of agricultural residues

Other

Waste

Solid waste disposal on land

Wastewater handling

Waste incineration

Other

ANNEX B

Party	Quantified emission limitation or reduction commitment (percentage of base year or period)
Australia	108
Austria	92
Belgium	92
Bulgaria	92
Canada	94
Croatia	95
Czech Republic	92
Denmark	92
Estonia	92
European Community	92
Finland	92
France	92
Germany	92
Greece	92
Hungary	94
Iceland	110
Ireland	92
Italy	92
Japan	94
Latvia	92
Liechtenstein	92
Lithuania	92
Luxembourg	92
Monaco	92
Netherlands	92
New Zealand	100
Norway	101
Poland	94

Portugal	92
Romania	92
Russian Federation	100
Slovakia	92
Slovenia	92
Spain	92
Sweden	92
Switzerland	92
Ukraine	100
United Kingdom of Great Britain and Northern Ireland	92
United States of America	93

Countries that are undergoing the process of transition to a market economy.

APPENDIX II

CLIMATE CHANGE: POLICY & REGIME

1989 Noordwijk Declaration

1. The composition of the earth's atmosphere is being seriously altered at an unprecedented rate due to human activity. Based on our current understanding, society is being threatened by man-made changes to the global climate.

2. While there are still uncertainties regarding the magnitude, timing and regional effects of climate change due to human activity, there is a growing consensus in the scientific community that significant climate change and instability are most likely over the next century.

Predictions available today indicate potentially severe economic and social dislocations for future generations.

Assuming these predictions, delay in action may endanger the future of the planet as we know it.

3. Fortunately, there is a growing awareness among the world population and their political leaders that action is needed. The basic principle of ecologically sustainable development has gained wide currency following the report of the World Commission on Environment and Development. This principle should be fundamental to efforts to tackle the problem of climate change and atmospheric pollution. The protection of the ozone layer is being addressed by the 1985 Vienna Convention on the Protection of the Ozone Layer and the 1987 Montreal Protocol on Substances that Deplete the Ozone Layer. Further strengthening of control measures contained in the Protocol was called for at the London Conference on Saving the Ozone Layer in March 1989 and the first meeting of the parties to the Montreal

Protocol at Helsinki in May 1989. A decision will be taken by the second meeting of the contracting parties to be held in London in 1990. The process aims at phasing out the production and consumption of Chlorofluorocarhons (CFCs) controlled under the Montreal Protocol by the year 2000 by the developed countries. They should also phase out other controlled substances which deplete the ozone layer as soon as feasible. Developing countries should also phase out these substances as soon as possible after their technology and resource needs are met.

4. Global warming is being addressed by the Intergovernmental Panel on Climate Change (IPCC), which was established by UNEP and WMO, and recognised by UN General Assembly Resolution 43/53 on Protection of global climate for present and future generations of mankind. The Hague Declaration of March 1989 put forward challenging ideas for international co-operation and legal and institutional measures. The 15th session of the UNEP Governing Council and the XLI session of the WMO Executive Council in 1989 requested their executive heads to begin preparations for negotiations on a framework convention on climate; these negotiations should be initiated as soon as possible after the interim report of the IPCC is adopted. This interim report will be reviewed at the Second World Climate Conference in November 1990. The 1989 Economic Summit agreed that a framework convention on climate change setting out general principles was urgently required and that specific protocols containing concrete commitments could be fitted into the framework as scientific evidence requires and permits. The Economic Summit also strongly advocated common efforts to limit emissions of carbon dioxide and other greenhouse gases. The July 1988 declaration of the states, parties to the Warsaw Treaty and the meeting non-aligned countries in Belgrade in September

1989 also addressed the issue of climate change. The Tokyo Conference on Global Environment and Human Response Towards Sustainable Development was held in September 1989. The Langkawi Declaration on Environment issued by the Common-wealth Heads of Governments in October 1989 stated the need to take new action to address the serious deterioration in the environment, including climate change. Given this base it is now time for governments of all countries to commit themselves to the IPCC, to strengthen and to extend the process of addressing climate change.

5. Measures to limit climate change will have other significant benefits such as reducing acidification, protecting the ozone layer, preserving biodiversity and other natural resources, preventing mean sea-level change and promoting sustainable development.

6. The Conference recognises the principle of the sovereign right of States to manage their natural resources independently. The Conference also reaffirms that global environmental problems have to be approached through international co-operation. Solving the external debt problem of developing countries, and establishing fair economic and commercial relationships between industrialised and developing countries would assist developing countries in creating appropriate conditions to protect the environment.

7. Climate change is a common concern of mankind. All countries should now, according to their capabilities and the means at their disposal, initiate actions and develop and maintain effective and operational strategies to control, limit or reduce emissions of greenhouse gases. As a first step, they should take those actions which are beneficial in their own right. Industrialised countries, in view of their contribution to the increase of greenhouse gas concentrations,

and in view of their capabilities, have specific responsibilities of different kinds: i) they should set an example by initiating domestic action, ii) they should support, financially and otherwise, the action by countries to which the protection of the atmosphere and adjustment to climate change would prove to be an excessive burden and iii) they should reduce emissions of greenhouse gases, also taking into account the need of the developing countries to have sustainable development. Developing countries establishing industrial facilities for the first time have a unique opportunity to include up-to-date technologies for controlling the emissions of greenhouse gases.

8. For the long term safeguarding of our planet and maintaining its ecological balance, joint effort and action should aim at limiting or reducing emissions and increasing sinks for greenhouse gases to a level consistent with the natural capacity of the planet. Such a level should be reached within a time frame sufficient to allow ecosystems to adapt naturally to climate change, to ensure that food production is not threatened and permit economic activity to develop in a sustainable and environmentally sound manner. Stabilising the atmospheric concentrations of greenhouse gases is an imperative goal. The IPCC will need to report on the best scientific knowledge as to the options for containing climate change within tolerable limits. Some currently available estimates indicate that this could require a reduction of global anthropogenic greenhouse gas emissions by more than 50 per cent. These estimates should be the subject of further examination by the IPCC.

9. While striving to preserve the global environment it is important to work at the same time to ensure stable development of the world economy, in line with the concept of "sustainable development". Effort and action should

include: i) the phasing-out of CFCs controlled by the Montreal Protocol, which are responsible for about one fifth of projected global warming, by national action and international co-operation in the context of the Montreal Protocol. This includes financial assistance and transfer of technology and information. In this connection, it is important that the substitutes for CFCs also should not contribute significantly to the global warming problem, ii) action especially by industrialised countries to limit or reduce CO_2-emissions, iii) action to reduce deforestation, prevent soil erosion and desertification; increase afforestation, and sound forest management in the temperate as well as the tropical zones, iv) action to limit or reduce the emissions of all greenhouse gases other than CO_2 and their precursors and to increase the sinks for such substances and v) intensified efforts for technological breakthroughs, for example with regards to renewable energy and removal and re-utilisation of CO_2.

10. The conference recommends that appropriate fora, including the IPCC, consider the necessity and efficiency of the introduction of the concept of CO_2-equivalence. This would provide a single parameter to describe the radiative effects of the various greenhouse gases, including CFCs. Such a concept, after taking into account other environmental considerations creates a basis for negotiations in response measures for different greenhouse gases in the most cost-effective manner. The Conference further recommends the development of common definitions and the harmonisation of methods to calculate CO_2-emissions.

11. All countries should increase co-operation in developing new, environmentally sound technologies, to improve existing technologies and increasingly to use these technologies in order to limit climate change or adapt to it.

Maximum use should be made of existing international organisations, institutions and mechanisms, governmental and non-governmental, for technology co-operation with and transfer to interested countries, especially developing countries. Factors that impede effective transfer of appropriate technologies should be identified and measures implemented to overcome these impediments.

12. Progress in reducing atmospheric pollution depends not only on technical and economic issues but also on attitudinal and conceptual changes. All countries especially industrialised countries, should recognise the need to make their socio-economic activities and life-styles environmentally sound. Improved dissemination of information and better training of personnel is needed, both at the national and international level. Public awareness programmes including school curricula, should include the issue of climate change and its connection with the way individuals use energy and other natural resources. Wider public awareness can be supported by increased scientific evidence arising from systematic research and monitoring activities. The Conference calls upon the non-governmental organisations to participate, in co-operation with international, regional and national authorities, in the efforts that are needed to respond to the problems of global warming, more specifically in the field of education and awareness building.

13. Many countries, especially developing countries will require assistance in identifying the causes of anthropogenic climate change, in establishing its extent and effect and also in responding to it. They will need help in acquiring, using, developing and maintaining technologies that are appropriate to their industrial, energy, transport, forestry and agricultural infrastructure. Industrialised countries will take steps to facilitate the transfer to developing

countries of technologies to limit the global climate change through financial assistance and other mechanisms to overcome the incremental costs of acquiring and using these technologies. Furthermore, the capabilities of these countries should be increased so that they can develop appropriate technologies themselves. In this context the concept of assured access to appropriate technologies in relation to proprietory rights needs to be explored.

Given this Stage of Development of the Issue of Climate Change, the Conference More Specifically

Carbon Dioxide (CO_2)

14. *Urges* all countries to take steps individually and collectively, to promote better energy conservation and efficiency and the use of environmentally sound energy sources, practices and technologies with no or minimum environmentally damaging characteristics. These policies should be reflected in short and long term energy policies and be pursued by all relevant sectors, including industry and transport, taking into account the need of developing countries for an adaptation period in order to enable them to meet their technological and other developmental needs. One direct means of allowing markets to incorporate the risk of climate change could be to ensure that the prices of all fuels reflect their full social, long run marginal and environmental costs and benefits.

15. *Agrees* that it is timely to investigate quantitative emission targets to limit or reduce CO_2 emissions and encourages the IPCC, in their interim report due in 1990, to include an analysis of target options.

16. *Recognises* the need to stabilise, while ensuring stable development of the world economy, CO_2 emissions and emissions of other greenhouse gases not controlled by

the Montreal Protocol. Industrialised nations agree that such stabilisation should be achieved by them as soon as possible, at levels to be considered by the IPCC and the Second World Climate Conference of November 1990. In the view of many industrialised nations such stabilisation of CO_2 emissions should be achieved as a first step at the latest by the year 2000.

> *Urges* all industrialised countries to support the process of IPCC through the investigation of the feasibility of achieving targets to limit or reduce CO_2 emissions including e.g. a 20 per cent reduction of CO_2 emission levels by the year 2005 as recommended by the scientific World Conference on the Changing Atmosphere in Toronto 1988.
>
> *Urges* all industrialised countries to intensify their efforts in this respect, while ensuring sustainable development and taking into account the specific circumstances of individual countries.

17. *Agrees* that industrialised countries with as yet, relatively low energy requirements, which can reasonably be expected to grow in step with their development, may have targets that accommodate that development.

18. *Calls* on the IPCC to present the analysis and conclusions referred to above to the Second World Climate Conference in November 1990.

19. *Agrees* that developing countries endeavour to meet future targets for CO_2-emissions and sinks, with due regard to their development requirements and within the limits of their financial and technical capabilities. International co-operation, whenever available, would be a contributing factor for greater action. New processes or industries to be

introduced should, as far as possible, incorporate technologies which are more energy-efficient and produce less pollution than present technologies.

20. *Agrees* that developing countries will need to be assisted financially and technically, including assistance with training, i.e. by strengthening relevant mechanisms to ensure that they will be in a position to manage, develop, and conserve their forest resources in a sustainable and environmentally sound manner. This will also contribute to combatting erosion and desertification. Recognition by the market of the total value of forests, including non-wood values, is a precondition for developing countries' being able to successfully use such financial and technical assistance for sustainable forest management.

21. *Agrees* to pursue a global balance between deforestation on the one hand and sound forest management and afforestation on the other. A world net forest growth of 12 million hectares a year in the beginning of next century should be considered as a provisional aim.

> *Requests* the IPCC to consider the feasibility of achieving this aim. To this end; the world deforestation rate should be slowed *inter alia* through the suppression of acid rain and other pollutants and of fires and through the reduction of pressures on biota. Sound forest management practices should be encouraged and at the same time vigorous forestry programmes should be developed in both temperate and tropical zones; biological diversity should be maintained; strategies addressing climate change issues through forest management and afforestation should be integrated with strategies addressing

the sustainability of other forest based values resulting in full multiple-use plans where appropriate, but with due consideration of the people living in or dependent on forest land.

Welcomes the work of the Tropical Forestry Action Plan and the International Timber Trade Organisation in pursuit of these goals.

Chlorofluorocarbons (CFCs)

22. *Welcomes* the commitment of the industrialised countries to amend the Montreal Protocol and to phase out the production and consumption of controlled Chlorofluorocarbons by the year 2000, and of other controlled ozone depleting substances as soon as feasible.

Urges all countries to become Parties to the Vienna Convention for the Protection of the Ozone Layer and to the Montreal Protocol. To facilitate this broad participation suitable amendments of the Montreal Protocol should be considered urgently by the Parties to the Protocol.

Urges industrialised countries to use financial and other means to assist developing countries in phasing out their production and consumption of controlled substances as soon as possible, by providing them with sufficient means to enable them to meet their target date. The development of alternative technologies and products in developing countries should be promoted.

Other Greenhouse Gases

23. *Recommends* that the development and implementation of specific means of limiting the atmospheric concentrations of greenhouse gases other than CO_2 and

CFCs should be energetically pursued, taking into consideration the special situation of developing countries.

Ministerial Meeting

24. *Recognises* the need to convene a Ministerial Conference to review the interim report of the IPCC. The conference endorses the plan of the organisation by WMO, UNEP, UNESCO and ICSU of such a meeting as part of the Second World Climate Conference in November 1990.

Funding

25.*Recommends* that existing institutions for development and financial assistance including the Multilateral Development Banks, Bilateral Assistance Programmes, the relevant United Nations organisations and specialised agencies, and scientific and technological organisations should give greater attention to climate change issues within their environmental and other relevant programmes by providing expanded funding including concessional funding. In addition, regional and subregional co-operation should be reinforced and funded so as to address and implement the required action at that level.

26. *Recommends* that additional resources should, over time, be mobilised to help developing countries take the necessary measures to address climate change and that are compatible with their development requirements.

Further *recommends* that the scope of resources needed must be assessed. Such assessments should include *inter alia* country studies and the capabilities of existing institutions and mechanisms to meet the financing needs identified, similar to the approaches developed under the Montreal Protocol.

Further consideration should be given to the need for funding facilities including a clearinghouse mechanism and a possible new international fund and their relationship to existing funding mechanisms, both multilateral and bilateral. Such funding should be related to the implementation of a future climate convention and associated protocols. In the meantime the donor community is urged to provide assistance to developing countries to support actions addressing climate change.

27. *Recommends* that, initially, international funding be directed towards,

(i) funding of a CFC phase-out in developing countries in the context of the Montreal Protocol;

(ii) promoting efficient use of energy, including appropriate end use technologies, increasing the use of non-fossil fuels and switching to energy sources with lower greenhouse gas emissions, and the use of renewable energy sources;

(iii) increased financial support for forest protection and forest management improvement, for example through the Tropical Forestry Action Plan (TFAP), the Plan of Action to Combat Desertification, the International Tropical Timber Organisation (ITTO) and other relevant international organisations;

(iv) assisting developing countries in planning how to address problems posed by climate change;

(v) supporting developing countries to enable their participation in the IPCC process and the other international meetings on this subject;

(vi) conducting research and monitoring;

(vii) arranging for technology transfer to and technology development in developing countries;

(viii) promoting public awareness, education and institutional and manpower development.

The use of financial resources could subsequently be extended *inter alia* to major energy sources with little or no environmentally damaging characteristics and for steps to reduce other global man-made emissions of greenhouse gases.

Research And Monitoring

28. *Urges* all countries and relevant organisations to increase their climate change research and monitoring activities and to provide for adequate data bases on emissions. *Also urges* states to co-operate in, and provide increasing support for, international co-ordination of these activities building on international programmes such as the World Climate Programme and the International Geosphere-Biosphere Programme, and on the present roles of the UNEP, WMO, ICSU, IEA, UNESCO, IOC, and other competent international organisations and bodies. The enhancement and strengthening of operational aspects of their work should be examined.

Recommends that more research should be carried out by 1992 into the sources and sinks of the greenhouse gases other than CO_2 and CFCs, like methane (CH_4), nitrous oxide (N_2O) and tropospheric ozone (O_3), including further research on the effect of the ocean on the concentration of radiatively active gases in the atmosphere.

Climate Change Convention

29.

1. *Urges* all countries to join and intensify the ongoing work within UNEP and WMO through the IPCC with respect to the compilation of elements for a framework convention on climate change so that negotiations upon it can start as soon as possible after the adoption of the interim report of the IPCC.
2. *Recommends* that such convention will be framed in such a way as to gain the adherence of the largest possible number and most suitably balanced spread of countries.
3. *Agrees* that to this end the framework convention and associated protocols should commit the parties *inter alia* to:

   enhancement of research and systematic observation of climate, aimed at detecting and monitoring climate variations and change;

   action to deal with greenhouse gas emissions and the effects of global warming;

   address the particular financial needs of the developing countries in the access to and transfer of technology; and

   strengthen sustainable forest management.

4. *Agrees* further that in developing the framework convention on climate change special attention should be given to ensuring that provision is made for appropriate decision making procedures and powers.
5. *Urges* all involved or to be involved in the negotiations to do their utmost to conclude these negotiations to ensure adoption of the convension as early as 1991 if possible and no later than at the Conference of the United Nations on Environment and Development in 1992.
6. *Considers* that in the preparation of the framework convention and protocols the relevant aspects of the Vienna Convention on the Protection of the Ozone Layer should be taken into account, as well as innovative approaches may be required by the complex character of the problem.

30. *Recommends* that this declaration and the supporting papers be conveyed to the IPCC at the conclusion of this Conference for further consideration and action.

List of participating countries: Argentina; Australia; Austria; Bangladesh; Belgium; Bolivia; Brazil; Bulgaria; Cameroon; Canada; China; Colombia; Costa Rica; Cuba; Denmark; Egypt; Finland; France; Gambia; German Democratic Republic; Germany, Federal Republic of Ghana; Greece; Guinea; Hungary; India; Indonesia; Ireland; Israel; Ivory Coast; Italy; Jamaica; Japan; Jordon; Kenya; Luxembourg; Malavsia; Mali; Malta; Mexico; Netherlands; New Zealand; Nigeria; Norway; Pakistan; Philippines;

Poland; Portugal; Republic of Korea; Saudi Arabia; Senegal; Spain; Sweden; Switzerland; Thailand; Trinidad and Tobago; Tunisia; Turkey; United States of America; Union of Soviet Socialist Republics; United Kingdom of Great Britain and Northern Ireland; United Republic of Tanzania; Venezuela; Yugoslavia; Zaire; Zambia; Zimbabwe

Commission of European Communities and International Organisations

Commission of European Communities; Food and Agriculture Organisation; International Atomic Energy Agency; International Energy Agency; Organisation for Economic Cooperation and Development; United Nations Development Program; United Nations Education, Scientific and Cultural Organisation; United Nations Industrial Development Organisation; United Nations Environment Program; World Meteorological Organisation; World Bank.

APPENDIX III

CLIMATE CHANGE GUIDELINES

United Nations Framework Convention on Climate Change, Rio, June 1992

The Parties to this Convention,

Acknowledging that change in the Earth's climate and its adverse effects are a common concern of humankind,

Concerned that human activities have been substantially increasing the atmospheric concentrations of greenhouse gases, that these increases enhance the natural greenhouse effect, and that this will result on average in an additional warming of the Earth's surface and atmosphere and may adversely affect natural ecosystems and humankind,

Noting that the largest share of historical and current global emissions of greenhouse gases has originated in developed countries, that per capita emissions in developing countries are still relatively low and that the share of global emissions originating in developing countries will grow to meet their social and development needs,

Aware of the role and importance in terrestrial and marine ecosystems of sinks and reservoirs of greenhouse gases,

Noting that there are many uncertainties in predictions of climate change, particularly with regard to the timing, magnitude and regional patterns thereof,

Acknowledging that the global nature of climate change calls for the widest possible cooperation by all countries and their participation in an effective and appropriate international response, in accordance with their common but differentiated responsibilities and respective capabilities and their social and economic conditions,

Recalling the pertinent provisions of the Declaration of the United Nations Conference on the Human Environment, adopted at Stockholm on 16 June 1972,

Recalling also that States have, in accordance with the Charter of the United Nations and the principles of international law, the sovereign right to exploit their own resources pursuant to their own environmental and developmental policies, and the responsibility to ensure that activities within their jurisdiction or control do not cause damage to the environment of other States or of areas beyond the limits of national jurisdiction,

Reaffirming the principle of sovereignty of States in international cooperation to address climate change,

Recognising that States should enact effective environmental legislation,that environmental standards, management objectives and priorities should reflect the environmental and developmental context to which they apply, and that standards applied by some countries may be inappropriate and of unwarranted economic and social cost to other countries, in particular developing countries,

Recalling the provisions of General Assembly resolution 44/228 of 22 December 1989 on the United Nations Confrerence on Environment and Development, and resolutions 43/53 of 6 December 1988, 44/207 of 22 December 1989, 45/212 of 21 December 1990 and 46/169 of 19 December 1991 on protection of global climate for present and future generations of mankind,

Recalling also the provisions of General Assembly resolution 44/206 of 22 December 1989 on the possible adverse effects of sea level rise on islands and coastal areas, particularly low-lying coastal areas and the pertinent provisions of General Assembly resolution 44/172 of 19 December 1989 on the implementation of the Plan of Action to Combat Desertification,

Recalling further the Vienna Convention for the Protection of the Ozone Layer, 1985, and the Montreal Protocol on Substances that Deplete the Ozone Layer, 1987, as adjusted and amended on 29 June 1990,

Noting the Ministerial Declaration of the Second World Climate Conference adopted on 7 November 1990,

Conscious of the valuable analytical work being conducted by many States on climate change and of the important contributions of the World Meteorological Organisation, the United Nations Environment Programme and other organs, organisations and bodies of the United Nations system, as well as other international and intergovernmental bodies, to the exchange of results of scientific research and the coordination of research,

Recognising that steps required to understand and address climate change will be environmentally, socially and economically most effective if they are based on relevant scientific, technical and economic considerations and continually re-evaluated in the light of new findings in these areas,

Recognising that various actions to address climate change can be justified economically in their own right and can also help in solving other environmental problems,

Recognising also the need for developed countries to take immediate action in a flexible manner on the basis of clear priorities, as a first step towards comprehensive response strategies at the global, national and, where agreed, regional levels that take into account all greenhouse gases, with due consideration of their relative contributions to the enhancement of the greenhouse effect,

Recognising further that low-lying and other small island countries, countries with low-lying coastal, arid and semi-arid areas or areas liable to floods, drought and desertification, and developing countries with fragile mountainous ecosystems are particularly vulnerable to the adverse effects of climate change,

Recognising the special difficulties of those countries, especially developing countries, whose economies are particularly dependent on fossil fuel production, use and exportation, as a consequence of action taken on limiting greenhouse gas emissions,

Affirming that responses to climate change should be coordinated with social and economic development in an integrated manner with a view to avoiding adverse impacts on the latter, taking into full account the legitimate priority needs of developing countries for the achievement of sustained economic growth and the eradication of poverty,

Recognising that all countries, especially developing countries, need access to resources required to achieve sustainable social and economic development and that, in order for developing countries to progress towards that goal, their energy consumption will need to grow taking into account the possibilities for achieving greater energy efficiency and for controlling greenhouse gas emissions in general, including through the application of new technologies on terms which make such an application economically and socially beneficial,

Determined to protect the climate system for present and future generations,

Have agreed as follows:

Article 1

Definitions

For the purposes oif this Conventions:

1. "Adverse effects of climate change" means changes in the physical environment or biota resulting from climate change which have significant deleterious effects on the composition, resilience or productivity of natural and managed ecosystems or on the operation of socio-economic systems or on human health and welfare.
2. "Climate change" means a change of climate which is attributed directly or indirectly to human activity that alters the composition of the global atmosphere and which is in addition to natural climate variability observed over comparable time periods.
3. "Climate system" means the totality of the atmosphere, hydrosphere, biosphere and geosphere and their interactions.
4. "Emissions" means the release of greenhouse gases and/or their precursors into the atmosphere over a specified area and period of time.
5. "Greenhouse gases" means those gaseous constituents of the atmosphere, both natural and

anthropogenic, that absorb and re-emit infrared radiation.

6. "Regional economic integration organisation" means an organisation constituted by sovereign States of a given region which has competence in respect of matters governed by this Convention or its protocols and has been duly authorised, in accordance with its internal procedures, to sign, ratify, accept, approve or accede to the instruments concerned.
7. "Reservoir" means a component or components of the climate system where a greenhouse gas or a precursor of a greenhouse gas is stored.
8. "Sink" means any process, activity or mechanism which removes a greenhouse gas, an aerosol or a precursor of a greenhouse gas from the atmosphere.
9. "Source" means any process or activity which releases a greenhouse gas, an aerosol or a precursor of a greenhouse gas into the atmosphere.

Article 2

Objective

The ultimate objective of this Convention and any related legal instruments that the Conference of the Parties may adopt is to achieve, in accordance with the relevant provisions of the Convention, stabilisation of greenhouse gas concentrations in the atmosphere at a level that would prevent dangerous anthropogenic interference with the climate system. Such a level should be achieved within a time frame sufficient to allow ecosystems to adapt naturally to climate change, to ensure that food production is not threatened and to enable economic development to proceed in a sustainable manner.

Article 3

Principles

In their actions to achieve the objective of the Convention and to implement its provisions, the Parties shall be guided, *inter alia,* by the following:

1. The Parties should protect the climate system for the benefit of present and future generations of humankind, on the basis of equity and in accordance with their common but differentiated responsibilities and respective capabilities. Accordingly, the developed country Parties should take the lead in combating climate change and the adverse effects thereof.
2. The specific needs and special circumstances of developing country Parties, especially those that are particularly vulnerable to the adverse effects of climate change, and of those Parties, especially developing country Parties, that would have to bear a disproportionate or abnormal burden under the Convention, should be given full consideration.
3. The Parties should take precautionary measures to anticipate, prevent or minimise the causes of climate change and mitigate its adverse effects. Where there are threats of serious or irreversible damage, lack of full scientific certainty should not be used as a reason for postponing such measures, taking into account that policies and measures to deal with climate change should be cost-effective so as to ensure global benefits at

the lowest possible cost. To achieve this, such policies and measures should take into account different socio-economic contexts, be comprehensive, cover all relevant sources, sinks and reservoirs of greenhouse gases and adaptation, and comprise all economic sectors. Efforts to address climate change may be carried out cooperatively by interested Parties.

4. The Parties have a right to, and should, promote sustainable development. Policies and measures to protect the climate system against human-induced change should be appropriate for the specific conditions of each Party and should be integrated with national development programmes, taking into account that economic development is essential for adopting measures to address climate change.

5. The Parties should cooperate to promote a supportive and open international economic system that would lead to sustainable economic growth and development in all Parties, particularly developing country Parties, thus enabling them better to address the problems of climate change. Measures taken to combat climate change, including unilateral ones, should not constitute a means of arbitrary or unjustifiable discrimination or a disguised restriction on international trade.

Article 4

Comments

1. All Parties, taking into account their common but differentiated responsibilities and their specific national

and regional development priorities, objectives and circumstances, shall:

(a) Develop, periodically update, publish and make available to the Conference of the Parties, in accordance with Article 12, national inventories of anthropogenic emissions by sources and removals by sinks of all greenhouse gases not controlled by the Montreal Protocol, using comparable methodologies to be agreed upon by the Conference of the Parties;

(b) Formulate, implement, publish and regularly update national and, where appropriate, regional programmes containing measures to mitigate climate change by addressing anthropogenic emissions by sources and removals by sinks of all greenhouse gases not controlled by the Montreal Protocol, and measures to facilitate adequate adaptation to climate change;

(c) Promote and cooperate in the development, application and diffusion, including transfer, of technologies, practices and processes that control, reduce or prevent anthropogenic emissions of greenhouse gases not controlled by the Montreal Protocol in all relevant sectors, including the energy, transport, industry, agriculture, forestry and waste management sectors;

(d) Promote sustainable management, and promote and cooperate in the conservation and enhancement, as appropriate, of sinks and reservoirs of all greenhouse gases not controlled by the Montreal Protocol, including biomass, forests and oceans as well as other terrestrial, coastal and marine ecosystems;

(e) Cooperate in preparing for adaptation to the impacts of climate change; develop and elaborate appropriate and integrated plans for coastal zone management, water resources and agriculture, and for the protection and rehabilitation of areas, particularly in Africa, affected by drought and desertification, as well as floods;

(f) Take climate change considerations into account, to the extent feasible, in their relevant social, economic and environmental policies and actions, and employ appropriate methods, for example impact assessments, formulated and determined nationally, with a view to minimising adverse effects on the economy, on public health and on the quality of the environment, of projects or measures undertaken by them to mitigate or adapt to climate change;

(g) Promote and cooperate in scientific, technological, technical, socio-economic and other research, systematic observation and development of data archives related to the climate system and intended to further the understanding and to reduce or eliminate the remaining uncertainties regarding the causes, effects, magnitude and timing of climate change and the economic and social consequences of various response strategies;

(h) Promote and cooperate in the full, open and prompt exchange of relevant scientific, technological, technical, socio-economic and legal information related to the climate system and climate change, and to the economic and social consequences of various response strategies;

(i) Promote and cooperate in education, training and public awareness related to climate change and encourage the widest participation in this process, including that of non-governmental organisations; and

(j) Communicate to the Conference of the Parties information related to implementation, in accordance with Article 12.

2. The developed country Parties and other Parties included in annex I commit themselves specifically as provided for in the following:

(a) Each of these Parties shall adopt nationalpolicies and take corresponding measures on the mitigation of climate change, by limiting its anthropogenic emissions of greenhouse gases and protecting and enhancing its greenhouse gas sinks and reservoirs. These policies and measures will demonstrate that developed countries are taking the lead in modifying longer-term trends in anthropogenic emissions consistent with the objective of the Convention, recognising that the return by the end of the present decade to earlier levels of anthropogenic emissions of carbon dioxide and other greenhouse gases not controlled by the Montreal Protocol would contribute to such modification, and taking into account the differences in these Parties' starting points and approaches, economic structures and resource bases, the need to maintain strong and sustainable economic growth, available technologies and other individual circumstances, as well as the need for equitable and appropriate contributions by each of these Parties to the global effort regarding that objective. These Parties may implement such policies and measures jointly with other Parties and may assist other Parties in contributing to the achievement of the objective of the Convention and, in particular, that of this subparagraph;

(b) In order to promote progress to this end, each of these Parties shall communicate, within six months of the entry into force of the Convention for it and periodically thereafter, and in accordance with Article 12, detailed information on its policies and measures referred to in subparagraph (a) above, as well as on its resulting projected anthropogenic emissions by sources and removals by sinks of greenhouse gases not controlled by the Montreal Protocol for the period referred to in subparagraph (a), with the aim of returning individually or jointly to their 1990 levels these anthropogenic emissions of carbon dioxide and other greenhouse gases not controlled by the Montreal Protocol. This information will be reviewed by the Conference of the Parties, at its first session and periodically thereafter, in accordance with Article 7;

(c) Calculations of emissions by sources and removals by sinks of greenhouse gases for the purposes of subparagraph (b) above should take into account the best available scientific knowledge, including of the effective capacity of sinks and the respective contributions of such gases to climate change. The Conference of the Parties shall consider and agree on methodologies for these calculations at its first session and review them regularly thereafter;

(d) The Conference of the Parties shall, at its first session, review the adequacy of subparagraphs (a) and (b) above. Such review shall be carried out in the light of the best available scientific information and assessment on climate change and its impacts, as well as relevant technical, social and economic information. Based on this review, the Conference of the Parties shall take appropriate action, which may include the adoption of amendments to the commitments in subparagraphs (a) and (b) above. The Conference of the Parties, at its first session, shall also take decisions regarding criteria for joint implementation as indicated in subparagraph (a) above. A second review of subparagraphs (a) and (b) shall take place not later than 31 December 1998, and thereafter at regular intervals determined by the Conference of the Parties, until the objective of the Convention is met;

(e) Each of these Parties shall :

(i) coordinate as appropriate with other such Parties, relevant economic and administrative instruments developed to achieve the objective of the Convention; and

(ii) identify and periodically review its own policies and practices which encourage activities that lead to greater levels of anthropogenic emissions of greenhouse gases not controlled by the Montreal Protocol than would otherwise occur;

(f) The Conference of the Parties shall review, not later than 31 December 1998, available information with a view to taking decisions regarding such amendments to the lists in annexes I and II as may be appropriate, with the approval of the Party concerned;

(g) Any Party not included in annex I may, in its instrument of ratification, acceptance, approval or accession, or at any time thereafter, notify the Depositary that it intends to be bound by subparagraphs (a) and (b) above. The Depositary shall inform the other signatories and Parties of any such notification.

3. The developed country Parties and other developed Parties included in annex II shall provide new and additional financial resources to meet the agreed full costs incurred by developing country Parties in complying with their obligations under Article 12, paragraph 1. They shall also provide such financial resources, including for the transfer of technology, needed by the developing country Parties to meet the agreed full incremental costs of implementing measures that are covered by paragraph 1 of this Article and that are agreed between a developing country Party and the international entity or entities referred to in Article 11, in accordance with that Article. The implementation of these commitments shall take into account the need for adequacy and predictability in the flow of funds and the importance of appropriate burden sharing among the developed country Parties.

4. The developed country Parties and other developed Parties included in annex II shall also assist the developing country Parties that are particularly vulnerable to the

adverse effects of climate change in meeting costs of adaptation to those adverse effects.

5. The developed country Parties and other developed Parties included in annex II shall take all practicable steps to promote, facilitate and finance, as appropriate, the transfer of, or access to, environmentally sound technologies and know-how to other Parties, particularly developing country Parties, to enable them to implement the provisions of the Convention. In this process, the developed country Parties shall support the development and enhancement of endogenous capacities and technologies of developing country Parties. Other Parties and organisations in a position to do so may also assist in facilitating the transfer of such technologies.

6. In the implementation of their commitments under paragraph 2 above, a certain degree of flexibility shall be allowed by the Conference of the Parties to the Parties included in annex I undergoing the process of transition to a market economy, in order to enhance the ability of these Parties to address climate change, including with regard to the historical level of anthropogenic emissions of greenhouse gases not controlled by the Montreal Protocol chosen as a reference.

7. The extent to which developing country Parties will effectively implement their commitments under the Convention will depend on the effective implementation by developed country Parties of their commitments under the Convention related to financial resources and transfer of technology and will take fully into account that economic and social development and poverty eradication are the first and overriding priorities of the developing country Parties.

8. In the implementation of the commitments in this Article, the Parties shall give full consideration to what actions are necessary under the Convention, including actions related to funding, insurance and the transfer of technology, to meet the specific needs and concerns of developing country Parties arising from the adverse effects of climate change and/or the impact of the implementation of response measures, especially on:

(a) Small island countries;

(b) Countries with low-lying coastal areas;

(c) Countries with arid and semi-arid areas, forested areas and areas liable to forest decay;

(d) Countries with areas prone to natural disasters;

(e) Countries with areas liable to drought and desertification;

(f) Countries with areas of high urban atmospheric pollution;

(g) Countries with areas with fragile ecosystems, including mountainous ecosystems;

(h) Countries whose economies are highly dependent on income generated from the production, processing and export, and/or on consumption of fossil fuels and associated energy-intensive products; and

(i) Land-locked and transit countries.

Further, the Conference of the Parties may take actions, as appropriate, with respect to this paragraph.

9. The Parties shall take full account of the specific needs and special situations of the least developed countries in their actions with regard to funding and transfer of technology.

10. The Parties shall, in accordance with Article 10, take into consideration in the implementation of the commitments of the Convention the situation of Parties, particularly developing country Parties, with economies that are vulnerable to the adverse effects of the implementation of measures to respond to climate change. This applies notably to Parties with economies that are highly dependent on income generated from the production, processing and export, and/or consumption of fossil fuels and associated energy-intensive products and/or the use of fossil fuels for which such Parties have serious difficulties in switching to alternatives.

Article 5

Research and Systematic Observation

In carrying out their commitments under Article 4, paragraph 1(g), the Parties shall:

(a) Support and further develop, as appropriate, international and intergovern-mental programmes and networks or organisations aimed at defining, conducting, assessing and financing research, data collection and systematic observation, taking into account the need to minimise duplication of effort;

(b) Support international and intergovernmental efforts to strengthen systematic observation and national scientific and technical research capacities and capabilities, particularly in developing countries, and to promote access to, and the exchange of, data and analyses thereof obtained from areas beyond national jurisdiction; and

(c) Take into account the particular concerns and

needs of developing countries and cooperate in improving their endogenous capacities and capabilities to participate in the efforts referred to in subparagraphs (a) and (b) above.

Article 6

Education, Training and Public Awareness

In carrying out their commitments under Article 4, paragraph 1(i), the Parties shall:

(a) Promote and facilitate at the national and, as appropriate, subregional and regional levels, and in accordance with national laws and regulations, and within their respective capacities:

(i) the development and implementation of educational and public awareness programmes on climate change and its effects;

(ii) public access to information on climate change and its effects;

(iii) public participation in addressing climate change and its effects and developing adequate responses; and

(iv) training of scientific, technical and managerial personnel.

(b) Cooperate in and promote, at the international level, and, where appropriate, using existing bodies:

(i) the development and exchange of educational and public awareness material on climate change and its effects; and

(ii) the development and implementation of

education and training programmes, including the strengthening of national institutions and the exchange or secondment of personnel to train experts in this field, in particular for developing countries.

Article 7

Conference of the Parties

1. A Conference of the Parties is hereby established.

2. The Conference of the Parties, as the supreme body of this Convention, shall keep under regular review the implementation of the Convention and any related legal instruments that the Conference of the Parties may adopt, and shall make, within its mandate, the decisions necessary to promote the effective implementation of the Convention. To this end, it shall:

(a) Periodically examine the obligations of the Parties and the institutional arrangements under the Convention, in the light of the objective of the Convention, the experience gained in its implementation and the evolution of scientific and technological knowledge;

(b) Promote and facilitate the exchange of information on measures adopted by the Parties to address climate change and its effects, taking into account the differing circumstances, responsibilities and capabilities of the Parties and their respective commitments under the Convention;

(c) Facilitate, at the request of two or more Parties, the coordination of measures adopted by them to address climate change and its effects, taking into account the differing circumstances, responsibilities and capabilities of the Parties and their respective commitments under the Convention.

(d) Promote and guide, in accordance with the objective and provisions of the Convention, the development and periodic refinement of comparable methodologies, to be agreed on by the Conference of the Parties, *inter alia*, for preparing inventories of greenhouse gas emissions by sources and removals by sinks, and for evaluating the effectiveness of measures to limit the emissions and enhance the removals of these gases;

(e) Assess, on the basis of all information made available to it in accordance with the provisions of the Convention, the implementation of the Convention by the Parties, the overall effects of the measures taken pursuant to the Convention, in particular environmental, economic and social effects as well as their cumulative impacts and the extent to which progress towards the objective of the Convention is being achieved;

(f) Consider and adopt regular reports on the implementation of the Convention and ensure their publication;

(g) Make recommendations on any matters necessary for the implementation of the Convention;

(h) Seek to mobilise financial resources in accordance with Article 4, paragraphs 3, 4 and 5, and Article 11;

(i) Establish such subsidiary bodies as are deemed necessary for the implementation of the Convention;

(j) Review reports submitted by its subsidiary bodies and provide guidance to them;

(k) Agree upon and adopt, by consensus, rules of procedure and financial rules for itself and for any subsidiary bodies;

(l) Seek and utilise, where appropriate, the services and cooperation of, and information provided by, competent international organisations and intergovernmental and non-governmental bodies; and

(m) Exercise such other functions as are required for the achievement of the objective of the Convention as well as all other functions assigned to it under the Convention.

3. The Conference of the Parties shall, at its first session, adopt its own rules of procedure as well as those of the subsidiary bodies established by the Convention, which shall include decision-making procedures for matters not already covered by decision- making procedures stipulated in the Convention. Such procedures may include specified majorities required for the adoption of particular decisions.

4. The first session of the Conference of the Parties shall be convened by the interim secretariat referred to in Article 21 and shall take place not later than one year after the date of entry into force of the Convention. Thereafter, ordinary sessions of the Conference of the Parties shall be held every year unless otherwise decided by the Conference of the Parties.

5. Extraordinary sessions of the Conference of the Parties shall be held at such other times as may be deemed necessary

by the Conference, or at the written request of any Party, provided that, within six months of the request being communicated to the Parties by the secretariat, it is supported by at least one-third of the Parties.

6. The United Nations, its specialised agencies and the International Atomic Energy Agency, as well as any State member thereof or observers thereto not Party to the Convention, may be represented at sessions of the Conference of the Parties as observers. Any body or agency, whether national or international, governmental or non-governmental, which is qualified in matters covered by the Convention, and which has informed the secretariat of its wish to be represented at a session of the Conference of the Parties as an observer, may be so admitted unless at least one-third of the Parties present object. The admission and participation of observers shall be subject to the rules of procedure adopted by the Conference of the Parties.

Article 8

Secretariat

1. A secretariat is hereby established.

2. The functions of the secretariat shall be:

(a) To make arrangements for sessions of the Conference of the Parties and its subsidiary bodies established under the Convention and to provide them with services as required;

(b) To compile and transmit reports submitted to it;

(c) To facilitate assistance to the Parties, particularly developing country Parties, on request, in the compilation and communication of information required in accordance with the provisions of the Convention;

(d) To prepare reports on its activities and present them to the Conference of the Parties;

(e) To ensure the necessary coordination with the secretariats of other relevant international bodies;

(f) To enter, under the overall guidance of the Conference of the Parties, into such administrative and contractual arrangements as may be required for the effective discharge of its functions; and

(g) To perform the other secretariat functions specified in the Convention and in any of its protocols and such other functions as may be determined by the Conference of the Parties.

3. The Conference of the Parties, at its first session, shall designate a permanent secretariat and make arrangements for its functioning.

Article 9

Subsidiary Body for Scientific and Technological Advice

1. A subsidiary body for scientific and technological advice is hereby established to provide the Conference of the Parties and, as appropriate, its other subsidiary bodies with timely information and advice on scientific and technological matters relating to the Convention. This body shall be open to participation by all Parties and shall be multidisciplinary. It shall comprise government representatives competent in the relevant field of expertise. It shall report regularly to the Conference of the Parties on all aspects of its work.

2. Under the guidance of the Conference of the Parties, and drawing upon existing competent international bodies, this body shall:

(a) Provide assessments of the state of scientific knowledge relating to climate change and its effects;

(b) Prepare scientific assessments on the effects of measures taken in the implementation of the Convention;

(c) Identify innovative, efficient and state-of-the-art technologies and know-how and advise on the ways and means of promoting development and/or transferring such technologies;

(d) Provide advice on scientific programmes, international cooperation in research and development related to climate change, as well as on ways and means of supporting endogenous capacity-building in developing countries; and

(e) Respond to scientific, technological and methodological questions that the Conference of the Parties and its subsidiary bodies may put to the body.

3. The functions and terms of reference of this body may be further elaborated by the Conference of the Parties.

Article 10

Subsidiary Body for Implementation

1. A subsidiary body for implementation is hereby established to assist the Conference of the Parties in the assessment and review of the effective implementation of the Convention. This body shall be open to participation by all Parties and comprise government representatives who are experts on matters related to climate change. It shall report regularly to the Conference of the Parties on all aspects of its work.

2. Under the guidance of the Conference of the Parties, this body shall:

(a) Consider the information communicated in accordance with Article 12, paragraph 1, to assess the overall aggregated effect of the steps taken by the Parties in the light of the latest scientific assessments concerning climate change;

(b) Consider the information communicated in accordance with Article 12, paragraph 2, in order to assist the Conference of the Parties in carrying out the reviews required by Article 4, paragraph 2(d); and

(c) Assist the Conference of the Parties, as appropriate, in the preparation and implementation of its decisions.

Article 11

Financial Mechanism

1. A mechanism for the provision of financial resources on a grant or concessional basis, including for the transfer of technology, is hereby defined. It shall function under the guidance of and be accountable to the Conference of the Parties, which shall decide on its policies, programme priorities and eligibility criteria related to this Convention. Its operation shall be entrusted to one or more existing international entities.

2. The financial mechanism shall have an equitable and balanced representation of all Parties within a transparent system of governance.

3. The Conference of the Parties and the entity or entities entrusted with the operation of the financial mechanism shall agree upon arrangements to give effect to

the above paragraphs, which shall include the following:

(a) Modalities to ensure that the funded projects to address climate change are in conformity with the policies, programme priorities and eligibility criteria established by the Conference of the Parties;

(b) Modalities by which a particular funding decision may be reconsidered in light of these policies, programme priorities and eligibility criteria;

(c) Provision by the entity or entities of regular reports to the Conference of the Parties on its funding operations, which is consistent with the requirement for accountability set out in paragraph 1 above; and

(d) Determination in a predictable and identifiable manner of the amount of funding necessary and available for the implementation of this Convention and the conditions under which that amount shall be periodically reviewed.

4. The Conference of the Parties shall make arrangements to implement the above mentioned provisions at its first session, reviewing and taking into account the interim arrangements referred to in Article 21, paragraph 3, and shall decide whether these interim arrangements shall be maintained. Within four years thereafter, the Conference of the Parties shall review the financial mechanism and take appropriate measures.

5. The developed country Parties may also provide and developing country Parties avail themselves of, financial resources related to the implementation of the Convention through bilateral, regional and other multilateral channels.

Article 12

Communication of Information Related to Implementation

1. In accordance with Article 4, paragraph 1, each Party shall communicate to the Conference of the Parties, through the secretariat, the following elements of information:

- (a) A national inventory of anthropogenic emissions by sources and removals by sinks of all greenhouse gases not controlled by the Montreal Protocol, to the extent its capacities permit, using comparable methodologies to be promoted and agreed upon by the Conference of the Parties;
- (b) A general description of steps taken or envisaged by the Party to implement the Convention; and
- (c) Any other information that the Party considers relevant to the achievement of the objective of the Convention and suitable for inclusion in its communication, including, if feasible, material relevant for calculations of global emission trends.

2. Each developed country Party and each other Party included in annex I shall incorporate in its communication the following elements of information:

- (a) A detailed description of the policies and measures that it has adopted to implement its commitment under Article 4, paragraphs 2(a) and 2(b); and
- (b) A specific estimate of the effects that the policies and measures referred to in subparagraph (a) immediately above will have on anthropogenic emissions by its sources and removals by its sinks of greenhouse gases during the period referred to in Article 4, paragraph 2(a).

3. In addition, each developed country Party and each other developed Party included in annex II shall incorporate details of measures taken in accordance with Article 4, paragraphs 3, 4 and 5.

4. Developing country Parties may, on a voluntary basis, propose projects for financing, including specific technologies, materials, equipment, techniques or practices that would be needed to implement such projects, along with, if possible, an estimate of all incremental costs, of the reductions of emissions and increments of removals of greenhouse gases, as well as an estimate of the consequent benefits.

5. Each developed country Party and each other Party included in annex I shall make its initial communication within six months of the entry into force of the Convention for that Party. Each Party not so listed shall make its initial communication within three years of the entry into force of the Convention for that Party, or of the availability of financial resources in accordance with Article 4, paragraph 3. Parties that are least developed countries may make their initial communication at their discretion. The frequency of subsequent communications by all Parties shall be determined by the Conference of the Parties, taking into account the differentiated timetable set by this paragraph.

6. Information communicated by Parties under this Article shall be transmitted by the secretariat as soon as possible to the Conference of the Parties and to any subsidiary bodies concerned. If necessary, the procedures for the communication of information may be further considered by the Conference of the Parties.

7. From its first session, the Conference of the Parties shall arrange for the provision to developing country Parties

of technical and financial support, on request, in compiling and communicating information under this Article, as well as in identifying the technical and financial needs associated with proposed projects and response measures under Article 4. Such support may be provided by other Parties, by competent international organisations and by the secretariat, as appropriate.

8. Any group of Parties may, subject to guidelines adopted by the Conference of the Parties, and to prior notification to the Conference of the Parties, make a joint communication in fulfilment of their obligations under this Article, provided that such a communication includes information on the fulfilment by each of these Parties of its individual obligations under the Convention.

9. Information received by the secretariat that is designated by a Party as confidential, in accordance with criteria to be established by the Conference of the Parties, shall be aggregated by the secretariat to protect its confidentiality before being made available to any of the bodies involved in the communication and review of information.

10. Subject to paragraph 9 above, and without prejudice to the ability of any Party to make public its communication at any time, the secretariat shall make communications by Parties under this Article publicly available at the time they are submitted to the Conference of the Parties.

Article 13

Resolution of Questioning Regarding Implementation

The Conference of the Parties shall, at its first session, consider the establishment of a multilateral consultative process, available to Parties on their request, for the resolution of questions regarding the implementation of the Convention.

Article 14

Settlement of Disputes

1. In the event of a dispute between any two or more Parties concerning the interpretation or application of the Convention, the Parties concerned shall seek a settlement of the dispute through negotiation or any other peaceful means of their own choice.

2. When ratifying, accepting, approving or acceding to the Convention, or at any time thereafter, a Party which is not a regional economic integration organisation may declare in a written instrument submitted to the Depositary that, in respect of any dispute concerning the interpretation or application of the Convention, it recognises as compulsory *ipso facto* and without special agreement, in relation to any Party accepting the same obligation:

(a) Submission of the dispute to the International Court of Justice, and/or

(b) Arbitration in accordance with procedures to be adopted by the Conference of the Parties as soon as practicable, in an annex on arbitration.

A Party which is a regional economic integration organisation may make a declaration with like effect in relation to arbitration in accordance with the procedures referred to in subparagraph (b) above.

3. A declaration made under paragraph 2 above shall remain in force until it expires in accordance with its terms or until three months after written notice of its revocation has been deposited with the Depositary.

4. A new declaration, a notice of revocation or the expiry of a declaration shall not in any way affect proceedings pending before the International Court of Justice or the arbitral tribunal, unless the parties to the dispute otherwise agree.

5. Subject to the operation of paragraph 2 above, if after twelve months following notification by one Party to another that a dispute exists between them, the Parties concerned have not been able to settle their dispute through the means mentioned in paragraph 1 above, the dispute shall be submitted, at the request of any of the parties to the dispute, to conciliation.

6. A conciliation commission shall be created upon the request of one of the parties to the dispute. The commission shall be composed of an equal number of members appointed by each party concerned and a chairman chosen jointly by the members appointed by each party. The commission shall render a recommendatory award, which the parties shall consider in good faith.

7. Additional procedures relating to conciliation shall be adopted by the Conference of the Parties, as soon as practicable, in an annex on conciliation.

8. The provisions of this Article shall apply to any related legal instrument which the Conference of the Parties may adopt, unless the instrument provides otherwise.

Article 15

Amendments to the Convention

1. Any Party may propose amendments to the Convention.

2. Amendments to the Convention shall be adopted at an ordinary session of the Conference of the Parties. The

text of any proposed amendment to the Convention shall be communicated to the Parties by the secretariat at least six months before the meeting at which it is proposed for adoption. The secretariat shall also communicate proposed amendments to the signatories to the Convention and, for information, to the Depositary.

3. The Parties shall make every effort to reach agreement on any proposed amendment to the Convention by consensus. If all efforts at consensus have been exhausted, and no agreement reached, the amendment shall as a last resort be adopted by a three-fourths majority vote of the Parties present and voting at the meeting. The adopted amendment shall be communicated by the secretariat to the Depositary, who shall circulate it to all Parties for their acceptance.

4. Instruments of acceptance in respect of an amendment shall be deposited with the Depositary. An amendment adopted in accordance with paragraph 3 above shall enter into force for those Parties having accepted it on the ninetieth day after the date of receipt by the Depositary of an instrument of acceptance by at least three-fourths of the Parties to the Convention.

5. The amendment shall enter into force for any other Party on the ninetieth day after the date on which that Party deposits with the Depositary its instrument of acceptance of the said amendment.

6. For the purposes of this Article, "Parties present and voting" means Parties present and casting an affirmative or negative vote.

Article 16

Adoption and Amendment of Annexes to the Convention

1. Annexes to the Convention shall form an integral

part thereof and, unless otherwise expressly provided, a reference to the Convention constitutes at the same time a reference to any annexes thereto. Without prejudice to the provisions of Article 14, paragraphs 2(b) and 7, such annexes shall be restricted to lists, forms and any other material of a descriptive nature that is of a scientific, technical, procedural or administrative character.

2. Annexes to the Convention shall be proposed and adopted in accordance with the procedure set forth in Article 15, paragraphs 2, 3, and 4.

3. An annex that has been adopted in accordance with paragraph 2 above shall enter into force for all Parties to the Convention six months after the date of the communication by the Depositary to such Parties of the adoption of the annex, except for those Parties that have notified the Depositary, in writing, within that period of their non-acceptance of the annex. The annex shall enter into force for Parties which withdraw their notification of non-acceptance on the ninetieth day after the date on which withdrawal of such notification has been received by the Depositary.

4. The proposal, adoption and entry into force of amendments to annexes to the Convention shall be subject to the same procedure as that for the proposal, adoption and entry into force of annexes to the Convention in accordance with paragraphs 2 and 3 above.

5. If the adoption of an annex or an amendment to an annex involves an amendment to the Convention, that annex or amendment to an annex shall not enter into force until such time as the amendment to the Convention enters into force.

Article 17

Protocols

1. The Conference of the Parties may, at any ordinary session, adopt protocols to the Convention.

2. The text of any proposed protocol shall be communicated to the Parties by the secretariat at least six months before such a session.

3. The requirements for the entry into force of any protocol shall be established by that instrument.

4. Only Parties to the Convention may be Parties to a protocol.

5. Decisions under any protocol shall be taken only by the Parties to the protocol concerned.

Article 18

Right to Vote

1. Each Party to the Convention shall have one vote, except as provided for in paragraph 2 below.

2. Regional economic integration organisations, in matters within their competence, shall exercise their right to vote with a number of votes equal to the number of their member States that are Parties to the Convention. Such an organisation shall not exercise its right to vote if any of its member States exercises its right, and vice versa.

Article 19

Depositary

The Secretary-General of the United Nations shall be the Depositary of the Convention and of protocols adopted in accordance with Article 17.

Article 20

Signature

This Convention shall be open for signature by States Members of the United Nations or of any of its specialised agencies or that are Parties to the Statute of the International Court of Justice and by regional economic integration organisations at Rio de Janeiro, during the United Nations Conference on Environment and Development, and thereafter at United Nations Headquarters in New York from 20 June 1992 to 19 June 1993.

Article 21

Interim Arrangements

1. The secretariat functions referred to in Article 8 will be carried out on an interim basis by the secretariat established by the General Assembly of the United Nations in its resolution 45/212 of 21 December 1990, until the completion of the first session of the Conference of the Parties.

2. The head of the interim secretariat referred to in paragraph 1 above will cooperate closely with the Intergovernmental Panel on Climate Change to ensure that the Panel can respond to the need for objective scientific and technical advice. Other relevant scientific bodies could also be consulted.

3. The Global Environment Facility of the United Nations Development Programme, the United Nations Environment Programme and the International Bank for Reconstruction and Development shall be the international entity entrusted with the operation of the financial mechanism referred to in Article 11 on an interim basis. In this connection, the Global Environment Facility should be appropriately

restructured and its membership made universal to enable it to fulfil the requirements of Article 11.

Article 22

Ratification, Acceptance, Approval or Accession

1. The Convention shall be subject to ratification, acceptance, approval or accession by States and by regional economic integration organisations. It shall be open for accession from the day after the date on which the Convention is closed for signature. Instruments of ratification, acceptance, approval or accession shall be deposited with the Depositary.

2. Any regional economic integration organisation which becomes a Party to the Convention without any of its member States being a Party shall be bound by all the obligations under the Convention. In the case of such organisations, one or more of whose member States is a Party to the Convention, the organisation and its member States shall decide on their respective responsibilities for the performance of their obligations under the Convention. In such cases, the organisation and the member States shall not be entitled to exercise rights under the Convention concurrently.

3. In their instruments of ratification, acceptance, approval or accession, regional economic integration organisations shall declare the extent of their competence with respect to the matters governed by the Convention. These organisations shall also inform the Depositary, who shall in turn inform the Parties, of any substantial modification in the extent of their competence.

Article 23

Entry into Force

1. The Convention shall enter into force on the ninetieth day after the date of deposit of the fiftieth instrument of ratification, acceptance, approval or accession.

2. For each State or regional economic integration organisation that ratifies, accepts or approves the Convention or accedes thereto after the deposit of the fiftieth instrument of ratification, acceptance, approval or accession, the Convention shall enter into force on the ninetieth day after the date of deposit by such State or regional economic integration organisation of its instrument of ratification, acceptance, approval or accession.

3. For the purposes of paragraphs 1 and 2 above, any instrument deposited by a regional economic integration organisation shall not be counted as additional to those deposited by States members of the organisation.

Article 24

Reservations

No reservations may be made to the Convention.

Article 25

Withdrawal

1. At any time after three years from the date on which the Convention has entered into force for a Party, that Party may withdraw from the Convention by giving written notification to the Depositary.

2. Any such withdrawal shall take effect upon expiry of one year from the date of receipt by the Depositary of the notification of withdrawal, or on such later date as may be specified in the notification of withdrawal.

3. Any Party that withdraws from the Convention shall be considered asalso having withdrawn from any protocol to which it is a Party.

Article 26

Authentic Texts

The original of this Convention, of which the Arabic, Chinese, English, French, Russian and Spanish texts are equally authentic, shall be deposited with the Secretary-General of the United Nations.

It Witness Whereof the undersigned, being duly authorised to that effect, have signed this Convention.

Done at New York this ninth day of May one thousand nine hundred and ninety-two.

GLOSSARY

Abrupt Climate Change: A change in climate over a widespread area that takes place so rapidly and unexpectedly that human and natural systems have difficulty adapting. An abrupt climate change occurs on the scale of decades, rather than centuries, and persists for years.

Afforestation: Planting of new forests on lands that have not been recently forested.

Albedo: Refers to the ratio of light from the sun that is reflected by the Earth's surface to the light received by it. Unreflected light is converted to infrared radiation (i.e., heat), which causes atmospheric warming (see "radiative forcing"). Thus, surfaces with a high albedo (e.g., snow and ice) generally contribute to cooling, whereas surfaces with a low albedo (e.g., forests) generally contribute to warming. Changes in land use that significantly alter the characteristics of land surfaces can therefore influence the climate through changes in albedo.

Alliance of Small Island States (AOSIS): A coalition of some 43 low-lying and small island countries, most of which are members of the G77, that are particularly vulnerable to the potential adverse consequences of climate change such as sea-level rise, coral bleaching, and increased frequency and intensity of tropical storms.

Allocation: Under an emissions trading scheme, permits to emit can initially either be given away for free, usually under a 'grandfathering' approach based on past emissions in a base year or an 'updating' approach based on the more recent emissions. The alternative is to auction permits in an initial market offering.

Ancillary Benefits: Complementary benefits of a climate policy including improvements in local air quality and reduced reliance of imported fossil fuels.

Assigned Amount: In the Kyoto Protocol, the permitted emissions, in CO_2 equivalents, during a commitment period. It is calculated using the Quantified Emission Limitation and Reduction Commitment (QELRC), together with rules specifying how and what emissions are to be counted.

Baselines: The baseline estimates of population, GDP, energy use and hence resultant greenhouse gas emissions without climate policies, determine how big a reduction is required, and also what the impacts of climate change without policy will be.

Base Year: Targets for reducing GHG emissions are often defined in relation to a base year. In the Kyoto Protocol, 1990 is the base year for most countries for the major GHGs; 1995 can be used as the base year for some of the minor GHGs.

Basket of Gases: This refers to the group six of greenhouse gases regulated under the Kyoto Protocol. They are listed in Annex A of the Kyoto Protocol and include: carbon dioxide (CO_2), methane (CH_4), nitrous oxide (N_2O), hydrofluorocarbons (HFCs), perfluorocarbons (PFCs), and sulphur hexafluoride (SF_6).

Berlin Mandate: Decision of the Parties reached at the first session of the Conference of the Parties to the UNFCCC (COP-1) in 1995 in Berlin that the commitments made by Annex I countries were inadequate and thus needed to be strengthened.

Black Carbon Aerosols: Particles of carbon in the atmosphere produced by inefficient combustion of fossil fuels or biomass. Black carbon aerosols absorb light from the sun, shading and cooling the Earth's surface, but contribute to significant warming of the atmosphere (see "radiative forcing").

Bryd-Hagel Resolution: In June 1997, anticipating the December 1997 meeting in Kyoto, Senator Robert C. Byrd (D-WV) introduced, with Sen. Chuck Hagel (R-NE) and 44 other cosponsors, a resolution stating that the impending Kyoto Protocol (or any subsequent international climate change agreement) should not -

Capital Stock: Existing investments in energy plant and equipment that may or may not be modified once installed.

Carbon Dioxide (CO_2): CO_2 is a colorless, odorless, non-poisonous gas that is a normal part of the ambient air. Of the six greenhouse gases normally targeted, CO_2 contributes the most to human-induced global warming. Human activities such as fossil fuel combustion and deforestation have increased atmospheric concentrations of CO_2 by approximately 30 percent since the industrial revolution. CO_2 is the standard used to determine the "global warming potentials" (GWPs) of other gases. CO_2 has been assigned a 100-year GWP of 1 (i.e., the warming effects over a 100-year time frame relative to other greenhouse gases).

Carbon Dioxide Equivalent (CO_2e): The emissions of a gas, by weight, multiplied by its "global warming potential."

Carbon Sinks: Processes that remove more carbon dioxide from the atmosphere than they release. Both the terrestrial biosphere and oceans can act as carbon sinks.

Carbon Taxes: A surcharge on the carbon content of oil, coal, and gas that discourages the use of fossil fuels and aims to reduce carbon dioxide emissions.

Certified Emissions Reduction (CER): Reductions of greenhouse gases achieved by a Clean Development Mechanism (CDM) project. A CER can be sold or counted toward Annex I countries' emissions commitments.

Reductions must be additional to any that would otherwise occur.

Chlorofluorocarbons (CFCs): CFCs are synthetic industrial gases composed of chlorine, fluorine, and carbon. They have been used as refrigerants, aerosol propellants, cleaning solvents and in the manufacture of plastic foam. There are no natural sources of CFCs. CFCs have an atmospheric lifetime of decades to centuries, and they have 100-year "global warming potentials" thousands of times that of CO_2, depending on the gas. In addition to being greenhouse gases, CFCs also contribute to ozone depletion in the stratosphere and are controlled under the Montreal Protcol.

Clean Development Mechanism (CDM): One of the three market mechanisms established by the Kyoto Protocol. The CDM is designed to promote sustainable development in developing countries and assist Annex I Parties in meeting their greenhouse gas emmissions reduction commitments. It enables industrialized countries to invest in emission reduction projects in developing countries and to receive credits for reductions achieved.

Climate: The long-term average weather of a region including typical weather patterns, the frequency and intensity of storms, cold spells, and heat waves. Climate is not the same as weather.

Climate change: Refers to changes in longe-term trends in the average climate, such as changes in average temperatures. In IPCC usage, climate change refers to any change in climate over time, whether due to natural variability or as a result of human activity. In UNFCCC usage, climate change refers to a change in climate that is attributable directly or indirectly to human activity that alters atmospheric composition.

Climate Sensitivity: The average global air surface temperature change resulting from a doubling of pre-industrial atmospheric CO_2 concentrations. The IPCC estimates climate sensitivity at 1.5-4.5°C (2.7-8.1°F).

Climate Variability: Refers to changes in patterns, such as precipitation patterns, in the weather and climate.

Commitment Period: The period under the Kyoto Protocol during which Annex I Parties' GHG emissions, averaged over the period, must be within their emission targets. The first commitment period runs from January 1, 2008 to December 31, 2012.

Conference of the Parties (COP): The supreme decision-making body comprised of the parties that have ratified the UN Framework Convention on Climate Change. It meets on an annual basis. As of February 2003, it is comprised of 188 countries.

Discounting: The process that reduces future costs and benefits to reflect the time value of money and the common preference of consumption now rather than later.

Early Crediting: A provision that allows crediting of emission reductions achieved prior to the start of a legally imposed emission control period. These credits can then be used to assist in achieving compliance once a legally imposed system begins.

Ecosystem: A community of organisms and its physical environment.

Emissions: The release of substances (e.g., greenhouse gases) into the atmosphere.

Emissions Cap: A mandated restraint in a scheduled timeframe that puts a "ceiling" on the total amount of anthropogenic greenhouse gas emissions that can be released into the atmosphere. This can be measured as gross emissions or as net emissions (emissions minus gases that are sequestered).

Emissions Reduction Unit (ERU): Emissions reductions generated by projects in Annex B countries that can be used by another Annex B country to help meet its commitments under the Kyoto Protocol. Reductions must be additional to those that would otherwise occur.

Emissions Trading: A market mechanism that allows emitters (countries, companies or facilities) to buy emissions from or sell emissions to other emitters. Emissions trading is expected to bring down the costs of meeting emission targets by allowing those who can achieve reductions less expensively to sell excess reductions (e.g. reductions in excess of those required under some regulation) to those for whom achieving reductions is more costly.

Energy Resources: The available supply and price of fossil and alternative resources will play a huge role in estimating how much a greenhouse gas constraint will cost. In the U.S. context, natural gas supply (and thus price) is particularly important, as it is expected to be a transition fuel to a lower carbon economy.

Enhanced Greenhouse Effect: The increase in the natural greenhouse effect resulting from increases in atmospheric concentrations of GHGs due to emissions from human activities.

Entry Into Force: The point at which international climate change agreements become binding. The United Nations Framework Convention on Climate Change (UNFCCC) has entered into force. In order for the Kyoto Protocol to do so as well, 55 Parties to the Convention must ratify (approve, accept, or accede to) the Protocol, including Annex I Parties accounting for 55 percent of that group's carbon dioxide emissions in 1990. As of June 2003, 110 countries had ratified the Protocol, representing 43.9 percent of Annex I emissions.

European Community: As a regional economic integration organization, the European Community can be and is a Party to the UNFCCC; however, it does not have a separate vote from its members (Austria, Belgium, Denmark, Finland, France, Germany, Greece, Ireland, Italy, Luxemburg, the Netherlands, Portugal, Spain, Sweden, and the United Kingdom).

Evapotranspiration: The process by which water re-enters the atmosphere through evaporation from the ground and transpiration by plants.

Global Warming: The progressive gradual rise of the Earth's average surface temperature thought to be caused in part by increased concentrations of GHGs in the atmosphere.

Global Warming Potential (GWP): A system of multipliers devised to enable warming effects of different gases to be compared. The cumulative warming effect, over a specified time period, of an emission of a mass unit of CO_2 is assigned the value of 1. Effects of emissions of a mass unit of non-CO_2 greenhouse gases are estimated as multiples. For example, over the next 100 years, a gram of methane (CH_4) in the atmosphere is currently estimated as having 23 times the warming effect as a gram of carbon dioxide; methane's 100-year GWP is thus 23. Estimates of GWP vary depending on the time-scale considered (e.g., 20-, 50-, or 100-year GWP), because the effects of some GHGs are more persistent than others.

Greenhouse Effect: The insulating effect of atmospheric greenhouse gases (e.g., water vapor, carbon dioxide, methane,etc.) that keeps the Earth's temperature about 60°F warmer than it would be otherwise.

Greenhouse Gas (GHG): Any gas that contributes to the "greenhouse effect."

Group of 77 and China, or G77/China: An international organization established in 1964 by 77 developing countries; membership has now increased to 133 countries. The group acts as a major negotiating bloc on some issues including climate change.

HGWP (High Global Warming Potential): Some industrially produced gases such as sulfur hexafluoride (SF_6), perfluorocarbons (PFCs), and hydrofluorocarbons (HFCs) have extremely high GWPs. Emissions of these gases have a much greater effect on global warming than an equal emission (by weight) of the naturally occurring gases. Most of these gases have GWPs 1,300—23,900 times that of CO_2. These GWPs can be compared to the GWPs of CO_2, CH_4, and N_2O which are presently estimated to be 1, 23 and 296, respectively.

"Hot Air": A situation in which emissions (of a country, sector, company or facility) are well below a target due to the target being above emissions that materialized under the normal course of events (i.e. without deliberate emission reduction efforts). Hot air can result from over-optimistic projections of growth. Emissions are often projected to grow roughly in proportion to GDP, and GDP is often projected to grow at historic rates. If a recession occurs and fuel use declines, emissions may be well below targets since targets are generally set in relation to emission projections. If emissions trading is allowed, an emitter could sell the difference between actual emissions and emission targets. Such emissions are considered hot air because they do not represent reductions from what would have occurred in the normal course of events.

Hydrofluorocarbons (HFCs): HFCs are synthetic industrial gases, primarily used in refrigeration and semi-conductor manufacturing as commercial substitutes for chlorofluorocarbons (CFCs). There are no natural sources of HFCs. The atmospheric lifetime of HFCs is decades to centuries, and they have 100-year "global warming potentials" thousands of times that of CO_2, depending on the gas. HFCs are among the six greenhouse gases to be curbed under the Kyoto Protocol.

Incentive-based Regulation: A regulation that uses the economic behavior of firms and households to attain desired environmental goals. Incentive-based programs involve taxes on emissions or tradable emission permits. The primary strength of incentive-based regulation is the flexibility it provides the polluter to find the least costly way to reduce emissions.

Intergenerational Equity: The fairness of the distribution of the costs and benefits of a policy when costs and benefits are borne by different generations. In the case of a climate change policy the impacts of inaction in the present will be felt in future generations.

Intergovernmental Panel on Climate Change (IPCC): The IPCC was established in 1988 by the World Meteorological Organization and the UN Environment Programme. The IPCC is responsible for providing the scientific and technical foundation for the United Nations Framework Convention on Climate Change (UNFCCC), primarily through the publication of periodic assessment reports (see "Second Assessment Report" and "Third Assessment Report").

Joint Implementation (JI): One of the three market mechanisms established by the Kyoto Protocol. Joint Implementation occurs when an Annex B country invests in an emissions reduction or sink enhancement project in another Annex B country to earn emission reduction units (ERUs).

Kyoto Mechanisms: The Kyoto Protocol creates three market-based mechanisms that have the potential to help countries reduce the cost of meeting their emissions reduction targets. These mechanisms are Joint Implementation (Article 6), the Clean Development Mechanisms (Article 12), and Emissions Trading (Article 17).

Kyoto Protocol: An international agreement adopted in December 1997 in Kyoto, Japan. The Protocol sets binding emission targets for developed countries that would reduce their emissions on average 5.2 percent below 1990 levels.

Land Use, Land-Use Change and Forestry (LULUCF): Land uses and land-use changes can act either as sinks or as emission sources. It is estimated that approximately one-fifth of global emissions result from LULUCF activities. The Kyoto Protocol allows Parties to receive emissions credit for certain LULUCF activities that reduce net emissions.

Market Benefits: Benefits of a climate policy that can be measured in terms of avoided market impacts such as changes in resource productivity (e.g., lower agricultural yields, scarcer water resources) and damages to human-built environment (e.g., coastal flooding due to sea-level rise).

Mauna Loa Record: The record of measurement of atmospheric CO_2 concentrations taken at Mauna Loa Observatory, Mauna Loa, Hawaii, since March 1958. This record shows the continuing increase in average annual atmospheric CO_2 concentrations.

Methane (CH_4): CH_4 is among the six greenhouse gases to be curbed under the Kyoto Protocol. Atmospheric CH_4 is produced by natural processes, but there are also substantial emissions from human activities such as landfills, livestock and livestock wastes, natural gas and petroleum systems, coalmines, rice fields, and wastewater treatment. CH_4 has a relatively short atmospheric lifetime of approximately 10 years, but its 100-year GWP is currently estimated to be approximately 23 times that of CO_2.

National Action Plans: Plans submitted to the Conference of the Parties (COP) by all Parties outlining the steps that they have adopted to limit their anthropogenic GHG emissions. Countries must submit these plans as a condition of participating in the UN Framework Convention on Climate Change and, subsequently, must communicate their progress to the COP regularly.

Negative Feedback: A process that results in a reduction in the response of a system to an external influence. For example, increased plant productivity in response to global warming would be a negative feedback on warming, because the additional growth would act as a sink for CO_2, reducing the atmospheric CO_2 concentration.

Nitrous Oxide (N_2O): N_2O is among the six greenhouse gases to be curbed under the Kyoto Protocol. N_2O is produced by natural processes, but there are also substantial emissions from human activities such as agriculture and fossil fuel combustion. The atmospheric lifetime of N_2O is approximately 100 years, and its 100-year GWP is currently estimated to be 296 times that of CO_2.

Non-Market Benefits: Benefits of a climate policy that can be measured in terms of avoided non-market impacts such as human-health impacts (e.g., increased incidence of tropical diseases) and damages to ecosystems (e.g., loss of biodiversity).

Perfluorocarbons (PFCs): PFCs are among the six types of greenhouse gases to be curbed under the Kyoto Protocol. PFCs are synthetic industrial gases generated as a by-product of aluminum smelting and uranium enrichment. They also are used as substitutes for CFCs in the manufacture of semiconductors. There are no natural sources of PFCs. PFCs have atmospheric lifetimes of thousands to tens of thousands of years and 100-year GWPs thousands of times that of CO_2, depending on the gas.

Positive Feedback: A process that results in an amplification of the response of a system to an external influence. For example, increased atmospheric water vapor in response to global warming would be a positive feedback on warming, because water vapor is a GHG and thus increases in water vapor in association with increases in greenhouse gases would cause greater warming than would occur if water vapor remained constant.

Radiative Forcing: The term radiative forcing refers to changes in the energy balance of the earth-atmosphere system in response to a change in factors such as greenhouse gas emissions, land-use change, or solar radiation. The climate system inherently attempts to balance incoming (e.g., light) and outgoing (e.g., heat) radiation. Positive radiative forcings increase the temperature of the lower atmosphere, which in turn increases temperatures at the Earth's surface. Negative radiative forcings cool the lower atmosphere. Radiative forcing is most commonly measured in units of watts per square meter (W/m^2).

Radiosondes: Sensors carried aboard weather balloons that have been in continuous use since 1979 for the monitoring of tropospheric temperatures.

Ratification: After signing the UNFCCC or the Kyoto Procotol, a country must ratify it, often with the approval of its parliament or other legislature. In the case of the Kyoto Protocol, a Party must deposit its instrument of ratification with the UN Secretary General in New York.

Reforestation: Replanting of forests on lands that have recently been harvested.

Regional Groups: The five regional groups meet privately to discuss issues and nominate bureau members and other officials. They are Africa, Asia, Central and Eastern Europe (CEE), Latin America and the Caribbean (GRULAC), and the Western Europe and Others Group (WEOG).

Renewable Energy: Energy obtained from sources such as geothermal, wind, photovoltaic, solar, and biomass.

Revenue Recycling: If permits are auctioned, this gives considerable sums of money to be recycled back into the economy, either through a lump sum payment of offsetting other taxes. If the existing taxes that are correspondingly reduced were very inefficient, this allows the possibility of both environmental and economic benefits from the trading system, commonly called the 'double dividend.'

Second Assessment Report (SAR): The Second Assessment Report, prepared by the Intergovernmental Panel on Climate Change, reviewed the existing scientific literature on climate change. Finalized in 1995, it is comprised of three volumes: Science; Impacts, Adaptations and Mitigation; and Economic and Social Dimensions of Climate Change.

Sequestration: Opportunities to remove atmospheric CO_2, either through biological processes (e.g. plants and trees), or geological processes through storage of CO_2 in underground reservoirs.

Sinks: Any process, activity or mechanism that results in the net removal of greenhouse gases, aerosols, or precursors of greenhouse gases from the atmosphere.

Stratosphere: The region of the Earth's atmosphere 10-50 km above the surface of the planet.

SRES Scenarios: A suite of emissions scenarios developed by the Intergovernmental Panel on Climate Change in its Special Report on Emissions Scenarios (SRES). These scenarios were developed to explore a range of potential future greenhouse gas emissions pathways over the 21st century and their subsequent implications for global climate change.

Subsidiary Body for Implementation (SBI): A permanent body established by the UNFCCC that makes recommendations to the COP on policy and implementation issues. It is open to participation by all Parties and is composed of government representatives.

Subsidiary Body for Scientific and Technological Advice (SBSTA): A permanent body established by the UNFCCC that serves as a link between expert information sources such as the IPCC and the COP.

Substitution: The economic process of trading off inputs and consumption due to changes in prices arising from a constraint on greenhouse gas emissions. How the extremely flexible U.S. economy adapts to available substitutes and/or finds new methods of production under a greenhouse gas constraint will be critical in minimizing overall costs of reducing emissions.

Sulfate Aerosols: Sulfur-based particles derived from emissions of sulfur dioxide (SO_2) from the burning of fossil fuels (particularly coal). Sulfate aerosols reflect incoming light from the sun, shading and cooling the

Earth's surface (see "radiative forcing") and thus offset some of the warming historically caused by greenhouse gases.

Sulfur Hexafluoride (SF_6): SF_6 is among the six types of greenhouse gases to be curbed under the Kyoto Protocol. SF_6 is a synthetic industrial gas largely used in heavy industry to insulate high-voltage equipment and to assist in the manufacturing of cable-cooling systems. There are no natural sources of SF_6. SF_6 has an atmospheric lifetime of 3,200 years. Its 100-year GWP is currently estimated to be 22,200 times that of CO_2.

Supplementarity: The Protocol does not allow Annex I parties to meet their emission targets entirely through use of emissions trading and the other Kyoto Mechanisms; use of the mechanisms must be supplemental to domestic actions to limit or reduce their emissions.

Targets and Timetables: Targets refer to the emission levels or emission rates set as goals for countries, sectors, companies, or facilities. When these goals are to be reached by specified years, the years at which goals are to be met are referred to as the timetables. In the Kyoto Protocol, a target is the percent reduction from the 1990 emissions baseline that the country has agreed to. On average, developed countries agreed to reduce emissions by 5.2% below 1990 emissions during the period 2008-2012, the first commitment period.

Technological Change: How much technological change will be additionally induced by climate policies is a crucial, but not well quantified, factor in assessing the costs of long-term mitigation of greenhouse gas emissions.

Thermal expansion: Expansion of a substance as a result of the addition of heat. In the context of climate change, thermal expansion of the world's oceans in response to global warming is considered the predominant driver of current and future sea-level rise.

Thermohaline Circulation (THC): A three-dimensional

pattern of ocean circulation driven by wind, heat and salinity that is an important component of the ocean-atmosphere climate system. In the Atlantic, winds transport warm tropical surface water northward where it cools, becomes more dense, and sinks into the deep ocean, at which point it reverses direction and migrates back to the tropics, where it eventually warms and returns to the surface. This cycle or "conveyor belt" is a major mechanism for the global transport of heat, and thus has an important influence on the climate. Global warming is projected to increase sea-surface temperatures, which may slow the THC by reducing the sinking of cold water in the North Atlantic. In addition, ocean salinity also influences water density, and thus decreases in sea-surface salinity from the melting of ice caps and glaciers may also slow the THC.

Third Assessment Report (TAR): The most recent Assessment Report prepared by the Intergovernmental Panel on Climate Change, which reviewed the existing scientific literature on climate change, including new information acquired since the completion of the Second Assessment report (SAR). Finalized in 2001, it is comprised of three volumes: Science; Impacts and Adaptation; and Mitigation.

Trace Gas: A term used to refer to gases found in the Earth's atmosphere other than nitrogen, oxygen, argon and water vapor. When this terminology is used, carbon dioxide, methane, and nitrous oxide are classified as trace gases. Although trace gases taken together make up less than one percent of the atmosphere, carbon dioxide, methane and nitrous oxide are important in the climate system. Water vapor also plays an important role in the climate system; its concentrations in the lower atmosphere vary considerably from essentially zero in cold dry air masses to perhaps 4 percent by volume in humid tropical air masses.

Troposphere: The region of the Earth's atmosphere 0-10 km above the planet's surface.

Umbrella Group: Negotiating group within the UNFCCC process comprising the United States, Canada, Japan, Australia, New Zealand, Norway, Iceland, Russia, and Ukraine.

Uncertainty: Uncertainty is a prominent feature of the benefits and costs of climate change. Decision makers need to compare risk of premature or unnecessary actions with risk of failing to take actions that subsequently prove to be warranted. This is complicated by potential irreversibilities in climate impacts and long term investments.

United Nations Framework Convention on Climate Change (UNFCCC): A treaty signed at the 1992 Earth Summit in Rio de Janeiro that calls for the "stabilization of greenhouse gas concentrations in the atmosphere at a level that would prevent dangerous anthropogenic interference with the climate system." The treaty includes a non-binding call for developed countries to return their emissions to 1990 levels by the year 2000. The treaty took effect in March 1994 upon ratification by more than 50 countries. The United States was the first industrialized nation to ratify the Convention.

Urban Heat Island (UHI): Refers to the tendency for urban areas to have warmer air temperatures than the surrounding rural landscape, due to the low albedo of streets, sidewalks, parking lots, and buildings. These surfaces absorb solar radiation during the day and release it at night, resulting in higher night temperatures.

Vector-borne disease: Disease that results from an infection transmitted to humans and other animals by blood-feeding anthropods, such as mosquitoes, ticks, and fleas. Examples of vector-borne diseases include Dengue fever, viral encephalitis, Lyme disease, and malaria.

Water Vapor (H_2O): Water vapor is the primary gas

responsible for the greenhouse effect. It is believed that increases in temperature caused by anthropogenic emissions of greenhouse gases will increase the amount of water vapor in the atmosphere, resulting in additional warming (see "positive feedback").

Weather: Describes the short-term (i.e., hourly and daily) state of the atmosphere. Weather is not the same as climate.

BIBLIOGRAPHY

Aagaard, K. and E. C. Carmack (1994). The Arctic Ocean and climate: a perspective. *The Polar Oceans and Their Role in Shaping the Global Environment*. 85: 5-20.

Alexander, V. (1998). Impacts of global change on Arctic marine ecosystems. *Global Change: Proceedings of the Symposium on the Arctic and Global Change*. J. McCullock. Washington, DC, Climate Institute.

Beardall, J., S. Beer, et al. (1998). "Biodiversity of marine plants in an era of climate change: Some predictions based on physiological performance." *Botanica Marina* 41(1): 113-123.

Bolin, R. L. and D. P. Abbott (1963). "Studies of the marine climate and phytoplankton of the central coastal area of California, 1954-1960." *CalCOFI* 9: 23-45.

Bugmann, H., A. Fischlin, et al. (1996). "Model convergence and state variable update in forest gap models." *Ecological Modelling* 89: 197-208.

Bugmann, H., R. Grote, et al. (1997). A new forest gap model to study the effects of environmental change on forest structure and functioning. *Proceedings of the Conference "Global Change Impacts on Tree Physiology and Forest Ecosystems*. G. M. J. Mohren and K. Kramer. Wageningen, The Netherlands, November, 1996., IBN-DLO, Wageningen, The Netherlands.

Bugmann, H., X. Yan, et al. (1996). "A comparison of forest gap models: Model structure and behaviour." *Climatic Change* 34: 289-313.

Buhlmann, B., P. Bossard, et al. (1987). "The influence of

longwave ultraviolet radiation (uv.-A) on the photosynthetic activity of phytoplankton." *Journal of Plankton Research* 9: 935-943.

Bush, M. B. (1996). "Amazonian conservation in a changing world." *Biological Conservation* 76(3): 219-228.

Buss, L. W. and J. R. Vaisnys (1993). "Temperature stress induces dynamical chaos in a cnidarian gastrovascular system." *Proceedings of the Royal Society of London Series B Biological Sciences* 252(1333): 39-41.

Butterfield, J. (1996). "Carabid life, cycle strategies and climate change: a study on an altitude transect." *Ecological Entomology* 21(1): 9-16.

Butterfield, J. E. and J. C. Coulson (1997). "Terrestrial invertebrates and climate change: physiological and life-cycle adaptations." *NATO ASI Series Series I Global Environmental Change* 47: 401-412.

Byrne, J. M. and A. A. Berg (1998). "Spatial analysis of strong El Nino event impacts on the Canadian prairies." *CWRJ* 23(3): 231-243.

Cadbury, J. (1995). "How might global warming affect the UK's flora and fauna?" *RSPB Conservation Review* 9: 48-52.

Cairns, C. (1997). Ultraviolet damaging fish DNA, says study. *The Scotsman*: 24.

Cayan, D. R. and D. H. Peterson, Eds. (1989). *The influence of North Pacific atmospheric circulation on streamflow.* Aspects of Climate Variability in the Pacific and Western Americas.

Center for Biodiversity and Conservation, A. M. o. N. H. (1999). Biodiversity and Climate Change: Conservation in the Face of Uncertainty (Conference proceedings).

Centre for Resource and Environmental Studies, I. o. A. S., Australian National University, Australia (1998). Climate change and the maintenance of conservation values in terrestrials ecosystems.

Chen, J. and J. F. Franklin (1997). "Growing-season microclimate variability within an old-growth Douglas-fir forest." *Climate Research* 8(1): 21-34.

Chown, S. and V. R. Smith (1993). "Climate change and the short-term impact of feral house mice at the sub-Antarctic Prince Edward Islands." *Oecologia* 96(4): 508-516.

Christian, K. A. and G. S. Bedford (1995). "Seasonal changes in thermoregulation by the frillneck lizard, Chlamydosaurus kingii, in tropical Australia." *Ecology* 76: 124-132.

Clark, J. S. (1988). "Forest dynamics, fire, and climate change during the last 2000 years at Itasca State Park, Northwestern Minnesota." *Bulletin of the Ecological Society of America Suppl.* 69(2): 100.

Clark, J. S. (1988). Climate change, fire occurrence, and forest influences during the last 750 years in Northwestern Minnesota. Minneapolis, Minnesota, University of Minnesota.

Clark, N. E., T. J. Blasing, et al. (1975). "Influence of inter-annual climatic fluctuations on biological systems." *Nature* 256: 302-305.

Clarke, A. (1990). Temperature and evolution: Southern Ocean Cooling and the Antarctic Marine Fauna. *Antarctic Ecosystems. Ecological Change and Conservation*. K. R. Kerry and G. Hempel. Berlin, Springer-Verlag: 9-22.

Clarke, A. (1996). "The influence of climate change on the distribution and evolution of organisms." *Society for Experimental Biology Seminar Series* 59: 377-407.

Clarke, A. and J. A. Crame (1993). The Southern Ocean benthic fauna and climate change: A historical perspective. *Antarctica and Environmental Change*. D. J. Drewry, R. M. Laws and J. A. Pyle. Oxford, Clarendon Press: 99-110.

Claussen, M. (1994). "On coupling global biome models with climate models." *Climate Research* 4: 203-221.

Claussen, M. (1996). "Variability of global biome patterns as a function of initial and boundary conditions in a climate model." *Climate Dynamics* 12: 371-379.

Claussen, M. (1997). "Modeling biogeophysical feedback in the African and Indian Monsoon region." *Climate Dynamics* 13: 247-257.

Claussen, M. (1998). "On multiple solutions of the atmosphere-vegetation system in present-day climate." *Global Change Biology* 4: 549-559.

Claussen, M. and V. Gayler (1997). "The greening of Sahara during the mid-Holocene: results of an interactive atmosphere—biome model." *Global Ecology and Biogeography Letters* 6: 369-377.

Claussen, M., C. Kubatzki, et al. (1999). "Simulation of an abrupt change in Saharan vegetation at the end of the mid-Holocene." *Geophysical Research Letters* 24(14): 2037-2040.

Claussen, M., V. Brovkin, et al. (1998). "Modeling global terrestrial vegetation—climate interaction." *Philosophical Trans. Royal Society of London* 353: 53-63.

Clifford, B. C., A. Davies, et al. (1996). "UK climate change models to predict crop disease and pest threats." *Aspects of Applied Biology* 45: 269-276.

climate and plant cover. Lesovedenie, v.0, n.2, 1999. March-April,:3-12." *Lesovedenie* 0(2): 3-12.

Closs, G. P. and P. S. Lake (1996). "Drought, differential mortality and the coexistence of a native and an introduced fish species in a south east Australian intermittent stream." *Environmental Biology of Fishes* 47: 17-26.

Cooney, R. T., T. M. Willette, et al. (1995). "The effect of climate on North Pacific pink salmon (Oncorhynchus gorbuscha) production: examining some details of a natural experiment." *Canadian Special Publication of Canadian Fisheries and Aquatic Sciences* 121: 475-482.

Coope, G. R. (1995). The effects of Quaternary climatic changes in insect populations: lessons from the past. *Insects in a Changing Environment*. R. Harrington and N. E. Stork. London, Academic Press.

Cronin, T. M. and C. E. Schneider (1990). "Climatic influences on species. Evidence from the fossil record." *Tends in Ecology and Evolution* 5(9): 275-279.

Dhakhwa, G. B. and C. L. Campbell (1998). "Potential effects of differential day-night warming in global climate change on crop production." *Climatic Change* 40: 647-667.

Diaz, R. A., G. O. Magrin, et al. (1997). "Climate change and its impact on the properties of agricultural soils in the Argentinean Rolling Pampas." *Climate Research* 9(1-2): 25-30.

Diaz, S. (1995). "Elevated CO_2 responsiveness, interactions at the community level and plant functional types." *Journal of Biogeography* 22(2-3): 289-295.

Diaz, S. and M. Cabido (1998). "Plant functional types and ecosystem function in relation to global change." *Journal of Vegetation Science* 8: 463-474.

Dickson, R., R. , K. R. Briffa, et al. (1994). "Cod and climate: the spatial and temporal context." *ICES Marine Science Symposia* 198(October): 280-286.

Dijkstra, P., et al. (1995). Interactions between atmospheric CO_2-concentration, temperature and environmental factors with respect to photosynthesis, assimilate distribution and development rhythm of three agricultural crops. Wageningen, The Netherlands, Institute for Agrobiology and Soil Fertility Research.

Dilley, M. (1998). "Climatic factors affecting annual maize yields in the valley of Oaxaca, Mexico." *International Journal of Climatology* 17(14): 1549-1558.

Dinar, A. (1998). *Measuring the impact of climate change on Indian agriculture*. Washington, D.C., World Bank.

Distributed in the USA by Routledge Chapman and Hall.

Distributors for the U.S. and Canada Elsevier Science Pub. Co.

Divoky, G. J. (1981). Birds and the ice-edge ecosystem in the Bering Sea. *The Eastern Bering Sea Shelf: Oceanography and Resources*. D. W. Hood and J. A. Calder. Seattle, Washington, University of Washington Press. 2.

Eaton, J. G. and R. M. Scheller (1996). "Effects of climate warming on fish thermal habitat instreams of the United States." *Limnol. Oceanogr.* 41(5): 1109-1115.

Ebbesmeyer, C. C. and R. M. Strickland (1995). Oyster Condition and Climate: Evidence from Willapa Bay. Seattle, Washington, Washington Sea Grant Program: 11 pp.

Ebbesmeyer, C. C., D. R. Cayan, et al. (1991). *1976 step in Pacific climate: Forty environmental changes between 1968-1975 and 1977-1984 . California Department of Water Resources*. Proceedings of the 7th Annual Pacific Climate (PACLIM) Workshop, April 1990.

EPA, U. S. (1995). Ecological impacts from climate change: an economic analysis of freshwater recreational fishing. Washington, DC, U.E. Environmental Protection Agency Policy, Planning and Evaluation Report.

EPA, U. S. (1995). Ecological Impacts from Climate Change: An Economic Analysis of Freshwater Recreational Fishing.,. Washington, DC.

Epstein, P. R., et. al (1998). "Biological and physical signs of climate change: Focus on mosquito-borne diseases." *Bulletin of the American Meteorological Society* 79: 409-417.

Ferrier, R. C., P. G. Whitehead, et al. (1993). "Potential impacts of afforestation and climate change on the stream water chemistry of the Monachyle catchment." *Journal of Hydrology* 145: 453-466.

Fiala, M. and L. Oriol (1990). "Light-temperature interactions on the growth of Antarctic diatoms." *Polar Biology* 10: 629-636.

Field, C. B., et. al (1999). Confronting Climate Change in California. Washington, DC, Union of Concerned Scientists

Field, C. D. (1995). "Impact of expected climate change on mangroves." *Hydrobiologia* 295(1-3): 75-81.

Fielding, C. A. (1999). "Predicting responses to climate change:

the effect of altitude and latitude on the phenology of the Spittlebug Neophilaenus lineatus." *Functional Ecology* 13(Supp. 1): 65-73.

Fosberg, M. A., et al. (1990). Global change: effects on forest ecosystems and wildfire severity. *Fire in the Tropical Biota. Ecosystem Processes and Global Challenges*. J. G. Goldammer. Berlin, Springer-Verlag. Ecological Studies, 84: 463-486.

Makadho, J. M. (1996). "Potential effects of climate change on corn production in Zimbabwe." *Climate Research* 6(2): 147-151. http://www.int-res.com/abstracts/cr/v06/n2/index.html.

Malcolm, J. R. and A. Markham (1997). Climate change threats to national parks and protected areas in the United States and Canada. Washington, D.C., World Wildlife Fund.

Malcolm, Jay R., and others. Global warming and extinctions of endemic species from biodiversity hotspots. *Conservation biology*, v. 20, Apr. 2006: 538-548.

Malcom, R. and C. Soulsby (1998). *The impact of climate change on a restored coastal wetland in north-east Scotland*. Second International Conference on Climate and Water, Espoo, Finland, Helsinki University of Technology.

Maley, J. (1996). "The African rain forest—main characteristics of changes in vegetation and climate from the Upper Cretaceous to the Quaternary." *Proceedings of the Royal Society of Edinburgh* 104-B: 31-73.

Malloy, K. D., M. A. Holman, et al. (1997). "Solar UVB-induced DNA damage and photoenzymatic DNA repair in Antarctic zooplankton." *Proceedings Of The National Academy Of Sciences Of The United States Of America* 94(4): 1258-1263.

Malmgren, B. A. and J. P. Kennett (1978). "Test size variation in *Globigerina bulloides* in response to Quaternary palaeoceanographic changes." *Nature* 275: 123-124.

Miller, K. A. and D. L. Fluharty (1992). El Nino and variability in the northeastern Pacific salmon fishery: implications for coping with climate change. *Climate variability, climate change and fisheries*. M. H. Glantz. Cambridge, New York, etc., Cambridge University Press: 49-88.

Mulchena, P. (1994). Implications of climate change for maize yield in Zimbabwe. *Implications of Climate Change for International Agriculture: Crop Modelling Study*. C. Rosenzweig and A. Inglesias. Washington, DC, U.S. Environmental Protection Agency.

Mulholland, P. J., C. R. Best, et al. (1997). "Effects of climate change on freshwater ecosystems of the Southeastern United States and the Gulf of Mexico." *Hydrological Processes* 11(8): 949-970.

Office, B. P. (1997). The impacts of global climate change in the Bering Sea Region: an assessment conducted by the International Arctic Science Commission under its Bering Sea Impact Study (BESIS). Fairbanks, Alaska, University of Alaska, Fairbanks: 40.

Poiani, K. A., C. W. Johnson, et al. (1995). "Sensitivity of Prairie Wetland to Increased Temperature and Seasonal Precipitation Changes." *Water Res. Bulletin* 31: 283-294.

Poladova, A. (1998). *Influence of climate to lake ecosystem of the Azerbaijan arid zones*. Second International Cofnerence on Climate and Water, Espoo, Finland, Helsinki University of Technology.

Polovina, J. J. (1996). "Decadal variation in the trans Pacific migration of northern bluefin tuna (Thunnus thynnus) coherent with climate-induced change in prey abundance." *Fisheries Oceanography* 5(2): 114-119.

Polovina, J. J. (1996). Climate and ecosystem variation in the North Pacific (abstract), 1996 Meeting of the International Whaling Commission.

Polovina, J., G. T. Mitchum, et al. (1993). "Physical and biological consequences of a climate event in the central North Pacific." *Fish. Ocean.* 3(1): 15.

Polson, S. W. (1997). "Mutations on the rise." *National parks* 71(5-6): 40.

Reilly, J. M. (1999). "Guest Editorial: Climate Change and Agriculture: The State of the Scientific Knowledge." *Climatic Change* 43(4): 645-650.

Reilly, J. M. and Schimmelpfenning (1999). "Agricultural impact assessment, vulnerability, and the scope for adaptation." *Climatic Change* 43(4): 745-788.

Retallack, S. (1999). "Wildlife in danger." *The Ecologist* 29(2): 102-103.

Revkin, Andrew C. Warming Causes Surge in Western Fires, Researchers Find. *The New York Times / Science / Environment*. July 7, 2006.

Risser, P. (1990). Impacts of climate change and variability on ecological systems. *The Rio Grande Basin: Global Climate Change Scenarios*. W. Stone, et al. Albuquerque, New Mexico, Water Resource Research Institute.

Roach, J. (1999). Rising temperatures threaten Antarctic ice shelves, nationalgeographic.com News, http://www.ngnews.com/news/1999/06/060299/ice_3478.asp.

Roberts, C. e. a. (1999). "The distribution of coral reef fish biodiversity: The climate-biodiversity connection." *Sea Wind* 12(4): 14-28.

Rogers, H. H., J. F. Thomas, et al. (1983). "Response of agronomic and forest species to elevated atmospheric carbon-dioxide." *Science* 220: 428-429.

Root, T. L. (1994). *Scientific / philosophical challenges of global change research: a case study of climatic changes on birds*. Proceedings of the American Philosophical Society.

Root, T. L. and S. H. Schneider (1993). "Can large-scale climatic models be linked with multiscale ecological studies?" *Conservation Biology* 7(2): 256-270.

Rose, C. and P. Hurst (1992). Can Nature Survive Global Warming? Gland, Switzerland, World Wide Fund for Nature: 59.

Rosenzweig, C. and A. Iglesias (1994). Implications of Climate Change for International Agriculture: Crop Modeling Study. Washington, DC, U.S. EPA.

Rosenzweig, C. and D. Hillel (1998). *Climate Change and the Global Harvest: Potential Impacts of the Greenhouse Effect on Agriculture*. New York, Oxford University Press.

Rosenzweig, C. and M. L. Parry (1994). "Potential impacts of climate change on world food supply." *Nature* 367: 133-138.

Rosenzweig, C., et al. (1993). Climate change and food supply. Oxford, U.K., University of Oxford, Environmental Change Unit.

Severnty, D. L. (1997). ". The use of data on the distribution of birds to monitor climatic changes." *Emu* 77: 162-166.

Shands, W. E., J. S. Hoffman, et al. (1987). *The Greenhouse effect, climate change, and U.S. forests*. Washington, D.C., Conservation Foundation.

Smith, R. C. and K. S. Baker (1980). "Stratospheric ozone, middle ultraviolet radiation, and carbon-14 measurements of marine productivity." *Science* 208: 592-593.

Smith, R. C., et al. (1992). "Ozone depletion: ultraviolet radiation and phytoplankton biology in Antarctic waters." *Science* 255: 952-959.

Wilkinson, C. R. and R. W. Buddemeier (1994). Global Climate Change and Coral Reefs: Implications for People and Reefs, UNEP-IOC-ASPEI-IUCN Global Task Team on the Implications of Climate Change on Coral Reefs: 122.

INDEX